A STUDENT'S ATLAS
OF FLOWERING PLANTS:
SOME DICOTYLEDONS
OF EASTERN NORTH AMERICA

Planned and Prepared Under the Direction of

CARROLL E. WOOD, JR.

Assisted by

ELIZABETH A. SHAW

With the Technical Help of

KAREN S. VELMURE and KENNETH R. ROBERTSON

Including 120 Illustrations Selected from the Work of
KAREN S. VELMURE, DOROTHY H. MARSH,
ARNOLD D. CLAPMAN, VIRGINIA SAVAGE, RACHEL A. WHEELER,
LAVERNE TRAUTZ, DIANE C. JOHNSON,
and SYDNEY B. DEVORE

HARPER & ROW, PUBLISHERS
New York, Evanston, San Francisco, London

INTRODUCTION

The 120 illustrations in this book and the lists of references to taxonomic terminology and to adaptations for pollination and seed dispersal are unanticipated results of work on a Generic Flora of the Southeastern United States which is being published in a series of papers in the <u>Journal</u> <u>of</u> <u>the</u> <u>Arnold</u> <u>Arboretum</u>. Since the publication of the first paper (treating genera of seven families), studies of some 89 additional families, as well as parts of the Compositae, Rosaceae, and Leguminosae, have been published. Treatments of some families are now in manuscript, those of others are approaching that stage. Preparation of illustrations based on fresh and preserved material has proceeded continuously, and drawings have been made for many genera for which treatments have not yet been prepared.

The principle stated in the introduction to the first paper that "illustrations which provide some insight into the details of the plant are far more desirable in a work at the generic level than a larger number of 'recognition drawings' " has been followed throughout the work. As a result, many details basic to the floral, dispersal, germination, and other systems of the biology of plants have been illustrated for the first time. For three years copies of some of the drawings have been used in the laboratory of the beginning taxonomy course at Harvard University. They have proved to be helpful to both student and professor. With this in mind, we have devised the system used here to provide the illustrations for a larger group of students and teachers, in the hope that all will benefit and will derive, from studying them, not only a clearer understanding of the components of the flora of eastern North America, but of flowering plants and their biology, wherever they occur.

The illustrations have been selected from among some 350 that have been prepared thus far for the Generic Flora to represent the basic families of dicotyledons in eastern North America. Because of their worldwide importance some families that are poorly represented in our flora (e. g., Myrtaceae) are included, along with several of intrinsic biological interest (e. g., Sarraceniaceae, Droseraceae, Rhizophoraceae). The wind-pollinated "Amentiferae" and a number of other genera with inconspicuous flowers that seldom are looked at closely are well represented, in the hope that more attention will be paid to their floral biology. In families such as Ranunculaceae, Rosaceae, Leguminosae, Ericaceae, and Compositae, a range of examples is included; those in the Rosaceae and Leguminosae represent each of the subfamilies. Illustrations are not yet available for some families that it would be desirable to include (e. g., Anacardiaceae), and our selection has had to be restricted to the dicotyledons because of gaps in our series of illustrations of monocotyledonous families.

In captioning the plates an attempt (perhaps futile) has been made to apply terminology uniformly across all the families. To avoid being overly repetitious a particular structure may not be labeled on every illustration in which it appears, but anyone who studies the drawings carefully will be able to interpret most of the families illustrated. As a further aid to the user lists of morphological and taxonomic terms and of adaptations for pollination and seed dispersal shown in the drawings are included.

The familial arrangement of the illustrations is approximately that of the Englerian system (as outlined by Diels in <u>A</u>. <u>Engler's</u> <u>Syllabus</u> <u>der</u> <u>Pflanzenfamilien</u>, ed. 11. 1936). In spite of the phylogenetic inadequacies of this classification, it is still the most completely worked out system; it is that used in most manuals and floras in all parts of the world; and it is that used in the arrangement of most major herbaria. From the standpoint of the student it is both convenient and logical, beginning with families in which the perianth is reduced or lacking, proceeding to those with a perianth of tepals, then to those with a calyx and a corolla of separate petals, and ending with the families in which the petals are united. Within each of these main groups families in which the ovary is superior generally precede those in which it is inferior. The scheme can be used rather like a synoptic key. It is useful as a framework on which the beginner can build; he can then proceed to classifications that are more assertedly phylogenetic (e. g., those of Takhtajan, Cronquist, and Thorne). For those who would rearrange the families and genera (or perhaps add notes), the illustrations have been printed on only one side of the page, and one might possibly take the book apart, punch the pages and interleave them with other material.

Since these illustrations were prepared for publication in another format, with a page approximately five inches wide and a reduction of approximately 50 per cent, the proportions and orientation are sometimes different on these larger pages where the amount of reduction ranges from about 12 to 35 per cent. In this larger size captions can be placed directly beside each figure in the illustration, but it has not been feasible to give the magnification of each component drawing.

The illustrations themselves are the work of eight different artists, whose initials accompany their drawings: Karen S. Velmure, Dorothy H. Marsh, Arnold D. Clapman, Virginia Savage, Rachel Ann Wheeler, LaVerne Trautz, Diane C. Johnston, and Sydney B. DeVore. Although each artist has a distinctive style, each has subordinated it to the overall requirement of a series of illustrations that are consistent with the style of drawing established by the first artist, the late Dorothy H. Marsh.

To make certain that the drawings are botanically accurate, all have been supervised, and most of the dissections for them have been prepared, either by me or by one or more of the fourteen botanists who have worked on the family and generic accounts in Cambridge. These botanists, in the order in which they worked on the Flora are R. B. Channell, C. W. James, K. A. Wilson, †W. R. Ernst, †G. K. Brizicky, S. A. Graham, I. K. Ferguson, A. L. Bogle, N. G. Miller, T. S. Elias, B. S. Vuilleumier, K. R. Robertson, S. A. Spongberg, and E. A. Shaw.

Dr. Shaw, who suggested preparing the captions for the illustrations on an IBM Selectric Composer, undertook the tedious job of positioning and typing all of the labels and of preparing the camera-ready copy for these pages and the index. Mrs. Velmure, who has drawn about a third of the plates, helped beyond measure with the time-consuming and tiring task of cutting out and attaching the labels and lines on about two-thirds of the illustrations and has added many details; I did the remainder, and I am responsible for the captions and the lists, including the inevitable errors. Dr. Robertson, who is working on the Generic Flora, supervised many of the illustrations that were made to fill in the gaps in our representation of families, prepared the plates for photography and mounted the prints, worked on the lists of terms and adaptations, and helped in other ways. In addition to having edited the Generic Flora accounts for ten years, Dr. Bernice G. Schubert has contributed her good advice and has very kindly checked the captions and other data.

The plant materials used in preparing the illustrations have come from many sources. Drawings are based either upon living plants, some grown at the Arnold Arboretum, or on preserved (alcoholic) specimens that we have collected or that have been sent through the kindness of many people. Those whose plant materials have been used in the illustrations in this book include George Avery, Ihsan Al Shehbaz, A. L. Bogle, †L. J. Brass, †G. K. Brizicky, Mrs. F. M. Carpenter, R. B. Channell, G. R. Cooley, F. C. Craighead, R. J. Eaton, T. S. Elias, I. K. Ferguson, A. J. Fordham, R. K. Godfrey, Alan Haney, †Mrs. J. Norman Henry, Josephine de N. Henry, R. A. Howard, Margaret Lefavour, Sidney McDaniel, A. B. Massey, N. G. Miller, L. I. Nevling, Jr., †H. F. L. Rock, R. C. Rollins, E. A. Shaw, T. J. Schultz, David Seligson, O. T. Solbrig, S. A. Spongberg, Alan Strahler, H. K. Svenson, J. W. Thieret, J. L. Thomas, R. M. and Alice F. Tryon, R. E. Umber, Richard Wagner, R. E. Weaver, R. L. Wilbur, K. A. Wilson, C. E. Wood, Jr., Elizabeth Wood, W. Woodrich, and R. L. Wyatt.

The work on the flora of the southeastern United States was stimulated by Dr. George R. Cooley, who has supported it generously. Much of the work on the Generic Flora has been supported by the National Science Foundation under Grants G2122, G9720, GB171, GB4111, and GB6459X (principal investigators, R. C. Rollins and/or C. E. Wood, Jr.). The Generic Flora has also had the continuing interest and support of Richard A. Howard, director of the Arnold Arboretum, and Reed C. Rollins, director of the Gray Herbarium, and that of our colleagues in both institutions and elsewhere during this time. We are grateful to all.

Carroll E. Wood, Jr.
Arnold Arboretum
Harvard University
Cambridge, Massachusetts
March, 1974

AN ATLAS OF DESCRIPTIVE TERMS

The terms listed below, with references to one or more genera and the pages where each is illustrated, are arranged more or less in the standard descriptive pattern for a flowering plant: habit, duration, roots, stems, leaves, inflorescences, flower type, flower parts, fruit, and seed. The terms have been selected to include many of those commonly used in taxonomic and morphological descriptions, with the expectation that illustrations will make terminology more understandable and meaningful. No attempt has been made to include all of the terms illustrated by these drawings, and some terms are not readily pointed out in a few words. Terms applying to vegetative parts of the plant are proportionately few in number, since these illustrations were prepared primarily to show reproductive structures and biology. The curious person will find that some terms in current use happen not to apply to any of the plants in this series, but he will find more than enough to stimulate study of the entire set of drawings in detail. Careful study of the illustrations will show that text-book definitions are more sharply defined than their real counterparts in living plants. Because of limitations of space a maximum of two lines of page references is devoted to each item, but in numerous instances the term is illustrated in many drawings.

LEAF

ARRANGEMENT

    alternate: Saururus, 2; Castanea, 12; Hamamelis, 50; Oxydendrum, 82; Calystegia, 92

    opposite: Pilea, 18; Arenaria, 26; Rhexia, 76; Scutellaria, 96; Diodia, 106

        decussate: Urtica, 17a; Chionanthus, 87h

    whorled: Casuarina, 1

    basal: Dodecatheon, 86; Tiarella, 47; Saxifraga, 48; Viola, 71

        basal & cauline: Silene, 27; Castilleia, 101

FORM

  PETIOLE

    absent:

        sessile: Arenaria, 26; Sedum, 46; Rhexia, 76; Scutellaria, 96a; Diodia, 106

        perfoliate: Triodanis, 113

    present, attached centrally (leaf peltate): Nelumbo, 28

    present, attached at base of blade: Saururus, 2; Salix, 3; Betula, 9; Tiarella, 47; Geranium, 62;
                    Passiflora, 72; Echinocystis, 111

        with glands near base of blade: Prunus, 54

        with a nectar gland: Albizia, 56

        with a pulvinus: Albizia, 56; Erythrina, 61

        winged: Poncirus, 63

  STIPULES

    free: Salix, 3o; Prunus, 54k; Cassia, 57m

    adnate to petiole: Rosa, 53a

    interpetiolar: Diodia, 106b

    ocrea: Polygonum, 22k

    spinescent: Erythrina, 61a

  BLADE

    form

        simple: Salix, 3; Populus, 4; Tiarella, 47; Passiflora, 72; Pedicularis, 100; Cichorium, 120

            entire: Magnolia, 34; Oenothera, 77; Cornus, 80; Chionanthus, 87; Mitchella, 107;

            palmately lobed: Delphinium, 32; Podophyllum, 33; Sanguinaria, 40; Geranium, 62; Acer, 68

            pinnately lobed: Comptonia, 5; Quercus, 13, 14; Stylophorum, 39; Salvia, 97k; Pedicularis, 100

        compound:

            palmate: Cannabis, 19; Potentilla, 52

compound blade (continued)
        even-pinnate (paripinnate): Cassia, 57
        odd-pinnate (imparipinnate): Carya, 6; Rosa, 53; Vicia, 60a
        bipinnate: Albizia, 56
        tripinnate: Daucus, 78a (lower leaves)
        trifoliolate: Baptisia, 59a; Erythrina, 61; Poncirus, 63
  venation
      palmate: Caltha, 29; Cercis, 58; Acer, 68; Kosteletzkya, 69; Echinocystis, 111; Sicyos, 112
      pinnate: Alnus, 8; Ostrya, 10; Quercus, 13; Ulmus, 15; Amaranthus, 24; Hamamelis, 50
      with strong secondary veins more or less paralleling primary vein: Pilea, 18g; Philadelphus, 49;
          Rhexia, 76; Plantago, 105

MODIFICATIONS

bracts:
    subtending flower: Casuarina, 1; Salix, 3; Comptonia, 5; Alnus, 8; Cannabis, 19; Lindera, 38;
        Calystegia, 92
      adnate to pedicel: Saururus, 2
      adnate to calyx: Verbena, 95c, f
    involucral:
      around single flower or 2 flowers (reduced inflorescence): Carya, 6a; Juglans, 7d; Fagus, 11e;
        Castanea, 12f; Quercus, 14d
      around a head of flowers: Compositae: Aster, 115; Helianthus, 117; Erechtites, 118; Cichorium, 120
      petaloid: Cornus, 80f
  scale-like: Casuarina, 1
  tendrils (modified leaflets): Vicia, 60a
  insect-trapping: Sarracenia (pitchers), 44; Drosera (tentacles), 45
  phyllodium: Sarracenia, 44

INFLORESCENCE

POSITION

solitary flower
    terminal: Asarum, 20a, h; Aristolochia, 21a (branching sympodial); Portulaca, 25i; Magnolia, 34;
        Liriodendron, 35; Sanguinaria, 40; Monotropa, 85a; Orobanche, 103a
    axillary: Carya, 6a (carpellate); Asimina, 36a, l; Oenothera, 77a; Ilex, 67d; Gaultheria, 83; Caly-
        stegia, 92
inflorescence
    terminal: Amaranthus, 24a, i; Capsella, 42a; Cruciferae, 43a, j, o; Pedicularis, 100a
    axillary: Amaranthus, 24a, i; Capsella, 42a; Drosera, 45; Rhizophora, 74; Hamamelis, 50

FORM

  raceme: Saururus, 2; Delphinium, 32a; Capsella, 42a; Selenia, 43j; Tiarella, 47; Prunus, 54a, c;
        Baptisia, 59; Croton, 65; Monotropa, 85k; Lobelia, 114
  spike: Melaleuca, 75; Verbena, 95o; Castilleia, 101; Plantago, 105
  corymb: Physocarpus, 51
  panicle: Saxifraga, 48a; Albizia, 56; Chionanthus, 87a; Oxydendrum, 82; Echinocystis, 111
  umbel: Maclura, 16b; Asclepias, 91c
    compound: Daucus, 78
  cyme: Arenaria, 26a; Silene, 27; Sedum, 46b; Potentilla, 52; Rhexia, 76; Sabatia, 88a
    helicoid: Drosera, 45
    scorpioid: Heliotropium, 93; Lithospermum, 94
    compound: Catalpa, 102
    with petaloid bracts: Cornus, 80f
  ament (catkin): Salix, 3; Comptonia, 5; Alnus, 8; Betula, 9; Ostrya, 10
  head: Casuarina, 1h; Maclura, 16d; Albizia, 56a; Cephalanthus, 108a; Compositae, 115, 117-120
  verticillate inflorescence: Casuarina, 1d, e; Salvia, 97a
  cyathium: Euphorbia & Chamaesyce, 66

FLOWER

FLOWER TYPE

complete: Arenaria, 26; Stylophorum, 39; Saxifraga, 48; Calystegia, 92; Lobelia, 114; Cichorium, 120

incomplete: Casuarina, 1; Salix, 3; Urtica, 17; Asarum, 20; Polygonum, 22; Echinocystis, 111; Ambrosia, 119

perfect ("hermaphroditic", "bisexual"): Polygonum, 22; Aquilegia, 31; Sanguinaria, 40; Lobelia, 114; Cichorium, 120

imperfect ("unisexual"): Casuarina, 1; genera on 3-14; Cannabis, 19; Amaranthus, 24; Ilex, 67; Acer, 68; Sicyos, 112; Amphiachyris, 115

plants monoecious (staminate & carpellate flowers on same plant): Alnus, 8; Betula, 9; Pilea, 18; Euphorbia, 66; Acer, 68; Ambrosia, 119

plants dioecious (staminate & carpellate flowers on different plants): Salix, 3; Populus, 4; Croton, 65; Ilex, 67

FLOWER PARTS

RECEPTACLE (torus): Nelumbo, 28; Magnolia, 34; Liriodendron, 35; Asimina, 36; Potentilla, 52

PERIANTH

symmetry

regular (actinomorphic, radially symmetrical): Asarum, 20; Arenaria, 26; Potentilla, 52; Albizia, 56; Geranium, 62; Calystegia, 92; Mitchella, 107

irregular (zygomorphic, bilaterally symmetrical): Aristolochia, 21; Delphinium, 32; Salvia, 97; Penstemon, 99; Cichorium, 120

insertion of perianth parts & stamens with respect to gynoecium

hypogynous: Arenaria, 26; Caltha, 29; Thalictrum, 30; Magnolia, 34; Asimina, 36; Poncirus, 63; Penstemon, 99

perigynous: Portulaca, 25c; Physocarpus, 51; Prunus, 54; Rhexia, 76

epigynous: Asarum, 20g; Opuntia, 73; Oenothera, 77; Daucus, 78; Helianthus, 117

aestivation of perianth parts (tepals, sepals, or petals)

valvate: Cannabis, 19d; Stylophorum, 39b (sepals); Sanguinaria, 40a (sepals); Philadelphus, 49b (sepals); Dodecatheon, 86b (sepals)

imbricate: Chenopodium, 23d; Nelumbo, 28a; Magnolia, 34; Prunus, 54c (petals); Pyrus, 55c (petals)

quincuncial: Polygonum, 22b; Dodecatheon, 86b, h (petals); Calystegia, 92a (sepals); Lithospermum, 94c (petals)

contorted: Amsonia, 89c; Calystegia, 92a (petals also plicate)

convolute: Philadelphus, 49b (petals); Oenothera, 77a, b (petals)

vexillary: Baptisia, 59a, b; Vicia, 60b, c; Erythrina, 61a, b

induplicate: Heliotropium, 93i, j

crumpled: Stylophorum, 39b (petals)

perianth absent: Casuarina, 1; Saururus, 2; Salix, 3; Comptonia, 5; Betula, 9f; Euphorbia & Chamaesyce, 66

perianth undifferentiated (composed of tepals)

apotepalous: Maclura, 16; Urtica, 17; Chenopodium, 23; Nelumbo, 28; Caltha, 29; Thalictrum, 30; Magnolia, 34; Opuntia, 73

syntepalous: Fagus, 11c; Ulmus, 15; Asarum, 20; Aristolochia, 21; Amaranthus, 24d; Acer, 68c, e

regular: Castanea, 12; Maclura, 16; Pilea, 18c; Asarum, 20; Caltha, 29

irregular: Pilea, 18d; Aristolochia, 21

accrescent (enlarging in fruit): Maclura, 16g; Urtica, 17h; Pilea, 18e, k; Polygonum, 22c

perianth of sepals & petals

calyx (sepals, collectively)

aposepalous (sepals free, distinct): Aquilegia, 31; Delphinium, 32; Liriodendron, 35; Asimina, 36; Capsella, 42; Sarracenia, 44

synsepalous (gamosepalous, connate): Silene, 27c; Vicia, 60b; Albizia, 56; Cercis, 58; Dodecatheon, 86; Pedicularis, 100

regular: Aquilegia, 31; Tiarella, 47; Philadelphus, 49; Rhizophora, 74; Dodecatheon, 86; Physalis, 98; Sambucus, 109

irregular: Cercis, 58; Vicia, 60; Scutellaria, 96; Pedicularis, 100; Castilleia, 101; Catalpa, 102

calyx (continued)
    bilabiate (2-lipped): Scutellaria, 96; Salvia, 97; Catalpa, 102
    with an epicalyx: Potentilla, 52; Kosteletzkya, 69
    petaloid: Aquilegia, 31; Delphinium, 32
    spurred: Delphinium, 32
    caducous: Stylophorum, 39; Sanguinaria, 40
    accrescent: Portulaca, 25f; Arenaria, 26f; Gaultheria, 83; Physalis, 98
corolla
    apopetalous (choripetalous, petals free, distinct): Portulaca, 25; Arenaria, 26; Sedum, 46;
        Baptisia, 59; Geranium, 62; Ilex, 67; Chimaphila, 84
    sympetalous (gamopetalous, petals connate): Rhododendron, 81; Amsonia, 89; Penstemon,
        99; Orobanche, 103; Lonicera,110; Erechtites, 118
    regular: Silene, 27; Aquilegia, 31; Potentilla, 52; Albizia, 56; Triodanis, 113
        salverform: Amsonia, 89; Heliotropium, 93h; Diodia, 106b; Mitchella, 107
        tubular: Albizia, 56b; Erigeron, 115b; Helianthus, 117e; Erechtites, 118
        funnelform (infundibuliform): Calystegia, 92; Physalis, 98; Diodia, 106f
        urceolate: Gaultheria, 83
        rotate: Sabatia, 88a, b
        reflexed: Dodecatheon, 86; Asclepias, 90a
    irregular: Delphinium, 32; Corydalis, 41; Saxifraga, 48; Pedicularis, 100; Lonicera, 110
        papilionaceous: Cercis, 58; Baptisia, 59; Vicia, 60; Erythrina, 61; Polygala, 64
        ligulate: Aster, 115a; Helianthus, 117c; Cichorium, 120
        two-lipped: 4 + 1: Scutellaria, 96; Lonicera, 110j
                2 + 3: Salvia, 97; Penstemon, 99; Pedicularis, 100; Castilleia, 101;
                    Catalpa, 102; Justicia, 104; Lobelia, 114m
        spurred (calcarate): Aquilegia, 31; Delphinium, 32
        salverform: Lithospermum, 94; Verbena, 95c
    with clawed petals: Silene, 27d
    with nectar glands: Corydalis, 41; Aquilegia, 31; Delphinium, 32; Lonicera, 110
    in two whorls: Liriodendron, 35; Asimina, 36

floral cup or floral tube ("hypanthium")
    perigynous nectar ring: Potentilla, 52
    floral cup: Tiarella, 47; Physocarpus, 51; Rosa, 53; Prunus, 54; Pyrus, 55
    floral tube: Rhexia, 76c; Oenothera, 77

ANDROECIUM (stamens, microsporophylls collectively)

    spirally arranged: Caltha, 29; Aquilegia, 31; Delphinium, 32; Magnolia, 34; Asimina, 36
    whorled: Chenopodium, 23; Silene, 27; Sassafras, 37; Tiarella, 47; Chimaphila, 84;
        Dodecatheon, 86
    free (distinct): Salix, 3; Arenaria, 26; Thalictrum, 30; Magnolia, 34; Stylophorum, 39
    synandrous (stamens united, connate) (see also under anther)
        monadelphous: Albizia, 56b; Polygala, 64; Kosteletzkya, 69; Sida, 70; Lobelia, 114
        diadelphous: Corydalis, 41; Vicia, 60; Erythrina, 61
        in fascicles or phalanges: Melaleuca, 75c
    tetradynamous (4 + 2): Capsella, 42
    irregular: Corydalis, 41; Cassia, 57; Cercis, 58; Erythrina, 61; Polygala, 64; Viola, 71; Rhexia, 76;
        Rhododendron, 81

anther & filament
    undifferentiated: Magnolia, 34
    hardly differentiated: Liriodendron, 35
    filament clavate: Thalictrum, 30j; Physalis, 98c

anthers
    sessile (filament absent): Juglans, 7; Viola, 71; Rhizophora, 74
    basifixed: Saururus, 2; Populus, 4; Podophyllum, 33d; Sanguinaria, 40; Poncirus, 63c

anthers (continued)

    dorsifixed: Sarracenia, 44c; Passiflora, 72; Cornus, 80b

        versatile: Passiflora, 72; Plantago, 105

    peltate: Albizia, 56

    connate (united, synantherous): Echinocystis, 111; Sicyos, 112; Erechtites, 118; Cichorium, 120

    with terminal appendages: Asclepias, 90b; Aster, 115g; Helianthus, 117f; Cichorium, 120f

    with basal appendages: Pterocaulon, 115h; Cichorium, 120f

    with nectariferous coronal hoods: Asclepias, 90

    with anther halves differentiated: Scutellaria, 96; Salvia, 97; Castilleia, 101; Justicia, 104

    divided into half anthers: Betula, 9 ; Ostrya, 10c, d; Carpinus, 10 l; Kosteletzkya, 69e

dehiscence of anthers

    longitudinal: Casuarina, 1e-g; Saururus, 2b; Alnus, 8c; Cannabis, 19e; Pyrus, 55e; Poncirus, 63; Dodecatheon, 86e

    by a slit across top or by chinks: Maclura, 16

    by flaps: Sassafras, 37; Lindera, 38; Hamamelis, 50

    by pores: Cassia, 57; Rhexia, 76; Rhododendron, 81; Oxydendrum, 82; Chimaphila, 84

    introrse (adaxial): Magnolia, 34c, l, m; Sassafras, 37; Lindera, 38; Triodanis, 113; Lobelia, 114; Compositae, 115, 117-120

    extrorse (abaxial): Asarum, 20d-f; Aristolochia, 21e, f; Podophyllum, 33c; Asimina, 36c, d

    lateral: Caltha, 29; Magnolia, 34p,q; Sassafras, 37

staminodia (nonfunctional, sterile, or reduced stamens): Aquilegia, 31; Sassafras, 37; Lindera, 38; Hamamelis, 50d, e; Acer, 68; Catalpa, 102; Penstemon, 99; Ilex, 67

pollen: Calystegia, 92; Cichorium, 120

    dimorphic: Lithospermum, 94

    in pollinia: Asclepias, 90

    with viscin strands: Oenothera, 77

adnation of androecium

    to corolla: Corydalis, 41g; Sida, 70b; Dodecatheon, 86; Sabatia, 88; Amsonia, 89; Calystegia, 92

    to gynoecium: Asarum, 20; Aristolochia, 21; Asclepias, 90

GYNOECIUM (carpels collectively)

    spirally arranged: Magnolia, 34; Liriodendron, 35

    whorled: Asarum, 20; Sedum, 46; Geranium, 62; Sida, 70

    apocarpous (carpels free, distinct): Caltha, 29; Thalictrum, 30; Asimina, 36; Potentilla, 52; Rosa, 53

    syncarpous (connate, united):

        by ovaries (and in some by styles up to stigmas): Saururus, 2c; Saxifraga, 48; Philadelphus, 49c, f; Calystegia, 92d; Rhododendron, 81c, h

        by styles and stigmas (ovaries free): Rosa, 53j; Amsonia, 89; Asclepias, 90

stigma

    lobed: Polygonum, 22b; Podophyllum, 33f; Stylophorum, 39c; Catalpa, 102; Lobelia, 114

    capitate: Capsella, 42b, c; Saxifraga, 48; Poncirus, 63c; Kosteletzkya, 69

    penicillate: Baptisia, 59; Pilea, 18d

    punctate: Tiarella, 47c, d, g; Potentilla, 52e; Erythrina, 61h

    decurrent: Ulmus, 15d, e; Thalictrum, 30o; Magnolia, 34

    terete: Alnus, 8h; Ostrya, 10e; Carpinus, 10n; Maclura, 16e

    funnelform: Monotropa, 85

    concave: Albizia, 56f

    with heteromorphic lobes (one sterile ?): Verbena, 95g; Scutellaria, 96h; Salvia, 97h, n

    sessile: Urtica, 17f; Ilex, 67f, g; Sambucus, 109

style

    gynobasic: Lithospermum, 94; Scutellaria, 96; Salvia, 97

    dimorphic (heterostylous): Lithospermum, 94; Mitchella, 107

style (continued):

    with a stylar canal: Saururus, 2c; Liriodendron, 35h, i; Sarracenia, 44; Monotropa, 85f

    with stylar brush or pollen-collecting hairs: Triodanis, 113; Lobelia, 114; Compositae, 116; Cichorium, 120

    with a stylopodium: Daucus, 78

    umbrella shaped (umbraculate): Sarracenia, 44

ovary

    position with respect to insertion of perianth & androecium

        superior: Ulmus, 15; Pilea, 18; Asarum, 20p; Poncirus, 63; Penstemon, 99
            with a gynophore: Warea, 43b
            with an androgynophore: Passiflora, 72e
        half-inferior (semi-inferior): Asarum, 20j; Portulaca, 25; Pyrus, 55; Rhexia, 76
        inferior: Castanea, 12; Asarum, 20g; Opuntia, 73; Oenothera, 77; Lobelia, 114

    placentation of ovules

        parietal (including marginal placentation of free carpels): Salix, 3; Podophyllum, 33; Stylophorum, 39; Capsella, 42; Drosera, 45; Passiflora, 72; Orobanche, 103
        axile: Aristolochia, 21; Poncirus, 63; Rhexia, 76; Rhododendron, 81; Physalis, 98
        axile below, parietal above: Sarracenia, 44; Monotropa, 85; Triodanis, 113
        free central: Arenaria, 26; Silene, 27; Dodecatheon, 86
        apical: Ulmus, 15; Thalictrum, 30o
        basal: Urtica, 17g; Polygonum, 22j; Portulaca, 25b, e; Calystegia, 92

    ovule:

        anatropous: Thalictrum, 30; Helianthus, 117f
        orthotropous: Saururus, 2c; Carya, 6; Urtica, 17g; Polygonum, 22j
        epitropous: Geranium, 62; Euphorbia, 66

INFRUCTESCENCE: see INFLORESCENCE

FRUIT & SEED

FRUIT

simple (from one carpel or from a single syncarpous gynoecium)

    pericarp (matured ovary wall inclosing seed or seeds), dry (see 5, 22, 53) or fleshy (see 33, 38, 67)

        achene: Saururus, 2; Alnus, 8m; Betula, 9j; Pilea, 18; Polygonum, 22; Thalictrum, 30; Potentilla, 52; Rosa, 53; Cichorium, 120

        nut: Comptonia, 5; Carya, 6; Juglans, 7; Fagus, 11; Quercus, 13, 14; Nelumbo, 28

        nutlet: an ambiguous term — see "achene" and "nut"; also see mericarps of 93 - 97 (cf. "schizocarp")

        utricle: Amaranthus, 24

        samara: Casuarina, 1; Liriodendron, 35k; Ulmus, 15k

        follicle: Caltha, 29; Aquilegia, 31; Delphinium, 32; Sedum, 46; Amsonia, 89; Asclepias, 91

        legume: Baptisia, 59; Vicia, 60; Erythrina, 61

        capsule: Salix, 3; Populus, 4; Stylophorum, 39; Corydalis, 41; Viola, 71; Castilleia, 101
            loculicidal: Philadelphus, 49; Oxydendrum, 82; Chimaphila, 84; Castilleia, 101
            septicidal: Philadelphus, 49; Rhododendron, 81
            poricidal: Triodanis, 113
            pyxis: Portulaca, 25e, f; Amaranthus, 24j; Plantago, 105

        silique (silicle): Cruciferae, 42, 43: replum, septum, valve, 43

        schizocarp (broadly defined as a fruit that splits into 1-seeded mericarps): Geranium, 62; Euphorbia, 66; Acer, 68; Daucus, 78; Umbelliferae, 79; Lithospermum, 94; Salvia, 97; Diodia, 106
            carpophore: Daucus, 78; Umbelliferae, 79
            columella: Croton, 65; Euphorbia, 66
            endocarp: Croton, 65; Diodia, 106

        berry: Podophyllum, 33h; Asimina, 36f; Opuntia, 73k; Lonicera, 110a
            with enlarged placenta: Physalis, 98j

        hesperidium: Poncirus, 63

        drupe:

            exocarp: Lindera, 38

drupe (continued):
    mesocarp: Lindera, 38
    endocarp (stone):
        1 endocarp: Maclura, 16; Sassafras, 37; Lindera, 38; Prunus, 54; Chionanthus, 87
        2 or more endocarps: Ilex, 67; Mitchella, 107; Sambucus, 109
    pome: Pyrus, 55

aggregate: Magnolia, 34 (follicles); Asimina, 36 (berries)

multiple (syncarp): Maclura, 16; Mitchella, 107 (both drupaceous)

accessory:
    hip: Rosa, 53
    accrescent perianth: Polygonum, 22; Amaranthus, 24e
    accrescent calyx: Gaultheria, 83
    accrescent bracts: Carpinus, 10; Ostrya, 10
    accrescent involucre: Ambrosia, 119

SEED

seed coat(s) (testa): Saururus, 2; Thalictrum, 30; Baptisia, 59; Ilex, 67; Opuntia, 73; Lonicera, 110
    sarcotesta (outer seed coat fleshy): Magnolia, 34; Passiflora, 72
    pleurogram: Albizia, 56
    micropyle: Caltha, 29
    hilum: Hamamelis, 50; Vicia, 60; Rhododendron, 81
    aril: Asarum, 20k; Stylophorum, 39; Sanguinaria, 40; Corydalis, 41; Polygala, 64h, i, o; Passi-
        flora, 72g, h
    caruncle: Croton, 65

aborted ovules: Casuarina, 1; Ostrya, 10j; Castanea, 12i; Quercus, 13g, 14h

endosperm
    present: Urtica, 17; Polygonum, 22; Caltha, 29; Physalis, 98; Penstemon, 99
    absent: Comptonia, 5; Quercus, 13; Sassafras, 37; Rosa, 53; Albizia, 56; Helianthus, 117
    perisperm: Saururus, 2

embryo
    plumule (epicotyl): Sassafras, 37; Lindera, 38; Baptisia, 59; Vicia, 60; Rhizophora, 74
    hypocotyl: Baptisia, 59; Vicia, 60; Rhizophora, 74; Calystegia, 92
    radicle: Sassafras, 37; Lindera, 38; Vicia, 60; Calystegia, 92
    types: (see A. C. Martin, American Midland Naturalist 36: 513 - 660. 1946)
        rudimentary: Caltha, 29i; Thalictrum, 30f; Aquilegia, 31n; Magnolia, 34h; Liriodendron, 35 l;
            Sanguinaria, 40i; Ilex, 67i
        broad: Saururus, 2g
        peripheral: Chenopodium, 23h, p; Amaranthus, 24h; Portulaca, 25h; Silene, 27j; Opuntia, 73m
        linear: Sarracenia, 44h; Daucus, 78m; Dodecatheon, 86 l
            curved: Physalis, 98j
        dwarf: Sedum, 46j; Saxifraga, 48i; Philadelphus, 49j; Penstemon, 99j
        spatulate: Casuarina, 1n; Comptonia, 5m; Urtica, 17j; Prunus, 54g, i; Polygala, 64i; Chionanthus,
            87j, k; Salvia, 97j; Helianthus, 117 l
            investing: Cercis, 58j; Poncirus, 63k; Echinocystis, 111i; Sicyos, 112g
            rudimentary: Mitchella, 107m
        bent: Urtica, 17j; Cannabis, 19k; Baptisia, 59m; Vicia, 60o; Erythrina, 61k; Justicia, 104g
            cotyledons incumbent: Capsella, 42h
            conduplicate (folded): Sinapis, 42g
            accumbent: Selenia, 42t; Justicia, 104g
        folded: Geranium, 62i, j; Acer, 68 l; Calystegia, 92k, l
        investing: Quercus, 13g; Ulmus, 15h; Sassafras, 37h; Lindera, 38i; Albizia, 56k
            with lobed cotyledons: Carya, 6,l; Juglans, 7i, l
            folded cotyledons: Fagus, 11

seed polyembryonic: Poncirus, 63

seedling: Carya, 6; Castanea, 12; Vicia, 60; Rhizophora, 74; Catalpa, 102; Erechtites, 118

SOME ADAPTATIONS FOR POLLINATION AND SEED DISPERSAL

Two of the critical points in the life cycle of flowering plants, the transfer of pollen from the stamen to the proper receptive stigma (pollination) and the removal of the seed from the parental sporophyte (dispersal), are reflected in a wide variety of morphological adaptations. Some of the more conspicuous examples that can be seen in the illustrations are listed below.

POLLINATION

MODES OF POLLINATION

Pollination by animal vectors: Although plants are assigned to particular categories below, few of them are completely restricted to any single vector, although some[*], e. g., Erythrina and Castilleia, are very nearly so. Many of our plants have no special adaptations to any single group of insects and are visited by members of several orders. Many more observations on pollinators need to be made.

insects:
beetles: Magnolia, 34
bees: Salix, 3; Liriodendron, 35; Sanguinaria, 40; Cassia, 57; Cercis, 58; Viola, 71; Rhexia, 76; Scutellaria, 96; Justicia, 104
mostly pollinated by long-tongued bees: Delphinium, 32; Baptisia, 59; Dodecatheon, 86; Salvia, 97; Penstemon, 99; Pedicularis, 100; Lobelia, 114m
butterflies: Heliotropium, 93h; Lithospermum, 94; Cephalanthus, 108
moths: Oenothera, 77; Lonicera, 110i, j
flies: Asimina 36 (flesh flies); Sassafras, 37; Euphorbia, 66a
various: Geranium, 62; Cornus, 80a; Asclepias, 90; Sambucus, 109; Helianthus, 117

birds (Archilochus colubris, ruby-throated hummingbird): Aquilegia, 31 (also visited by bees); Erythrina, 61; Castilleia, 101; Lonicera, 110b; Lobelia, 114a,b

pollination by wind: Casuarina, 1; genera on 4 - 16; Cannabis, 19; Chenopodium, 23; Amaranthus, 24; Thalictrum, 30 m-p; Acer, 68; Plantago, 105; Ambrosia, 119

pollination by rain: Caltha, 29 (also insect pollinated)

cleistogamous flowers (versus chasmogamous flowers): Polygala, 64a, k; Viola, 71m, n; Triodanis, 113b

NECTAR FLOWERS (except in Salix [dioecious] also offering pollen)

sites of nectar production
nectar disc or ring: Salix, 3; Philadelphus, 49; Physocarpus, 51; Prunus, 54; Erythrina, 61; Geranium, 62d; Poncirus, 63; Passiflora, 72e; Melaleuca, 75; Cornus, 80; Amsonia, 89; Calystegia, 92; Scutellaria, 96; Salvia, 97
glands on corolla: Lonicera, 110c
glands on stamens or glandular staminodia: Sassafras, 37; Lindera, 38; Viola, 71; Asclepias 90a, b (coronal hoods)
glands on gynoecium: Caltha, 29; Heliotropium, 93d
on spurred petals: Aquilegia, 31; Delphinium, 32; Corydalis, 41

POLLEN FLOWERS (nectar not produced): Stylophorum, 39; Sanguinaria, 40; Sarracenia, 44; Cassia, 57; Rhexia, 76; Sambucus, 109 (disc present but not nectariferous)

SOME MECHANISMS FOR OUTCROSSING

dioecism (plants dioecious): Salix, 3; Populus, 4; Croton, 65; Ilex, 67
protandry (proterandry): Ulmus, 15; Aquilegia, 31; Geranium, 62b; Sabatia, 88; Triodanis, 113; Lobelia, 114; Cichorium, 120
protogyny (proterogyny): Asarum, 20; Aristolochia, 21; Magnolia, 34; Plantago, 105
heterostyly: Lithospermum, 94; Mitchella, 107
anthesis of imperfect flowers at different times: Acer, 68a

FRUIT AND SEED DISPERSAL

DISPERSED BY VARIOUS ANIMALS

carried incidentally

fruit with awns, hooks, or spines: Sida, 70; Daucus, 78; Osmorhiza, 79m; Sanicula, 79q; Sicyos, 112
seed mucilaginous when wet (adaptation for dispersal or for germination ?): Urtica, 17; Plantago, 105h

DISPERSED BY VARIOUS ANIMALS (continued)

carried deliberately

fruit large, fleshy: Maclura, 16

seed large, carried by mammals or birds: Carya, 6; Castanea, 12; Quercus, 14

seed arillate, carried by ants: Asarum, 20; Stylophorum, 39; Sanguinaria, 40; Polygala, 64; Croton, 65 (carunculate)

fruit or seed eaten, seed passing through digestive tract

dispersal units dry

small achenes: Cannabis, 19; Polygonum, 22

small seeds: Portulaca, 25; Aquilegia, 31; Physocarpus, 51; Potentilla, 52; Calystegia, 92

seeds mimetic, conspicuous, bright red (bird dispersed): Erythrina, 61

dispersal units fleshy

calyx accrescent and fleshy: Gaultheria, 83

fruit fleshy with 1 or more endocarps: Sassafras, 37; Lindera, 38; Prunus, 54; Ilex, 67; Cornus, 80; Chionanthus, 87; Mitchella, 107; Sambucus, 109

pericarp fleshy, several seeded: Podophyllum, 33; Asimina, 36; Passiflora 72 (see below); Opuntia, 73; Physalis, 98

floral cup fleshy: Rosa, 53 (inclosing achenes): Pyrus, 55 (adnate to ovary)

outer seed coat fleshy (sarcotesta): Magnolia, 34; Passiflora, 72 (also arillate)

DISPERSED BY WIND

fruit with wing-like bracts: Carpinus & Ostrya, 10

fruit flattened: indehiscent legumes of Albizia, 56 and Cercis, 58; mericarps of Heracleum, 79a - c

fruit with wing(s): Casuarina, 1; Betula, 9; Ulmus, 15; Liriodendron, 35; Acer, 68; Oenothera, 77g, h

fruit with tuft of hairs (pappus) at apex: Erechtites, 118

seed tiny, often with tails or wings: Drosera, 45; Philadelphus, 49; Melaleuca, 75; Rhododendron, 81; Oxydendrum, 82; Chimaphila, 84; Monotropa, 85; Castilleia, 101; Orobanche, 103

seed larger, winged: Selenia, 43k; Oenothera, 77; Catalpa, 102

seed with tuft of hairs: Salix, 3; Populus, 4; Asclepias, 91

utricle with accrescent perianth: Chenopodium, 23; Amaranthus, 24

DISPERSED BY WATER

floating receptacle: Nelumbo, 28

floating fruit: Saururus, 2; Cephalanthus, 108

floating seed: Caltha, 29; Aristolochia, 21 (corky funiculus)

floating embryo: Rhizophora, 74

DISPERSED BALLISTICALLY

fruit opening explosively: Vicia, 60; Geranium, 62 (see below); Croton, 65; Euphorbia, 66; Justicia, 104

seed ejected from fruit by drying and consequent contraction of carpel wall: Hamamelis, 50; Geranium, 62 (see above); Viola, 71

achene ejected from receptacle by turgid staminodia beneath fruit: Pilea, 18

seeds or fruit ejected by falling raindrops: Tiarella, 47 (striking enlarged lower valve of capsule); Scutellaria, 96 (striking lower calyx lobe — upper lobe falls away earlier); Pedicularis, 100 (striking winglike sterile part of open capsule)

SYSTEMATIC LIST OF ILLUSTRATIONS

Family arrangement is that of Diels, A. Engler's Syllabus der Pflanzenfamilien, ed. 11. Berlin. 1936.

CASUARINACEAE: Casuarina, 1
SAURURACEAE: Saururus, 2
SALICACEAE: Salix, 3; Populus, 4
MYRICACEAE: Comptonia, 5
JUGLANDACEAE: Carya, 6; Juglans, 7
BETULACEAE: Alnus, 8; Betula, 9;
   Carpinus & Ostrya, 10
FAGACEAE: Fagus, 11; Castanea, 12;
   Quercus, 13, 14
ULMACEAE: Ulmus, 15
MORACEAE: Maclura, 16
URTICACEAE: Urtica, 17; Pilea, 18
CANNABACEAE: Cannabis, 19
ARISTOLOCHIACEAE: Asarum, 20;
   Aristolochia, 21
POLYGONACEAE: Polygonum, 22
CHENOPODIACEAE: Chenopodium, 23
AMARANTHACEAE: Amaranthus, 24
PORTULACACEAE: Portulaca, 25
CARYOPHYLLACEAE: Arenaria, 26;
   Silene, 27
NYMPHAEACEAE: Nelumbo, 28
RANUNCULACEAE: Caltha, 29; Thalictrum, 30;
   Aquilegia, 31; Delphinium, 32
BERBERIDACEAE: Podophyllum, 33
MAGNOLIACEAE: Magnolia, 34; Liriodendron, 35
ANNONACEAE: Asimina, 36
LAURACEAE: Sassafras, 37; Lindera, 38
PAPAVERACEAE: Stylophorum, 39;
   Sanguinaria, 40
FUMARIACEAE: Corydalis, 41
*CRUCIFERAE: Capsella & fruits, 42;
   fruits & seeds, 43
SARRACENIACEAE: Sarracenia, 44
DROSERACEAE: Drosera, 45
CRASSULACEAE: Sedum, 46
SAXIFRAGACEAE: Tiarella, 47;
   Saxifraga, 48; Philadelphus, 49
HAMAMELIDACEAE: Hamamelis, 50
ROSACEAE: Physocarpus, 51; Potentilla, 52;
   Rosa, 53; Prunus, 54; Pyrus, 55
*LEGUMINOSAE: Albizia, 56; Cassia, 57;
   Cercis, 58; Baptisia, 59; Vicia, 60;
   Erythrina, 61
GERANIACEAE: Geranium, 62
RUTACEAE: Poncirus, 63
POLYGALACEAE: Polygala, 64
EUPHORBIACEAE: Croton, 65; Euphorbia
   & Chamaesyce, 66

AQUIFOLIACEAE: Ilex, 67
ACERACEAE: Acer, 68
MALVACEAE: Kosteletzkya, 69; Sida, 70
VIOLACEAE: Viola, 71
PASSIFLORACEAE: Passiflora, 72
CACTACEAE: Opuntia, 73
RHIZOPHORACEAE: Rhizophora, 74
MYRTACEAE: Melaleuca, 75
MELASTOMATACEAE: Rhexia, 76
ONAGRACEAE: Oenothera, 77
*UMBELLIFERAE: Daucus, 78; fruit types, 79
CORNACEAE: Cornus, 80
ERICACEAE: Rhododendron, 81; Oxydendrum, 82;
   Gaultheria, 83; Chimaphila, 84; Monotropa, 85
PRIMULACEAE: Dodecatheon, 86
OLEACEAE: Chionanthus, 87
GENTIANACEAE: Sabatia, 88
APOCYNACEAE: Amsonia, 89
ASCLEPIADACEAE: Asclepias, 90, 91
CONVOLVULACEAE: Calystegia, 92
BORAGINACEAE: Heliotropium, 93; Lithospermum, 94
VERBENACEAE: Verbena, 95
*LABIATAE: Scutellaria, 96; Salvia, 97
SOLANACEAE: Physalis, 98
SCROPHULARIACEAE: Penstemon, 99; Pedicularis, 100; Castilleia, 101
BIGNONIACEAE: Catalpa, 102
OROBANCHACEAE: Orobanche, 103
ACANTHACEAE: Justicia, 104
PLANTAGINACEAE: Plantago, 105
RUBIACEAE: Diodia, 106; Mitchella, 107;
   Cephalanthus, 108
CAPRIFOLIACEAE: Sambucus, 109; Lonicera, 110
CUCURBITACEAE: Echinocystis, 111; Sicyos, 112
CAMPANULACEAE: Triodanis, 113; Lobelia, 114
*COMPOSITAE: inflorescence, flowers, & stamens, 115;
   stylar types, 116; Helianthus, 117; Erechtites, 118;
   Ambrosia, 119; Cichorium, 120

The names marked with an asterisk are those of families for which there is an alternative name. Either name is correct under Article 18 of the International Code of Botanical Nomenclature, 1972. Both names are used on the appropriate illustrations.

One-hundred nineteen genera in seventy families are more or less completely illustrated. The illustrations of fruits and seeds of Cruciferae, fruit types of Umbelliferae, and inflorescences, flowers, stamens, and stylar types of Compositae include representatives of thirty-seven additional genera.

CASUARINACEAE: Casuarina. a-n, C. equisetifolia

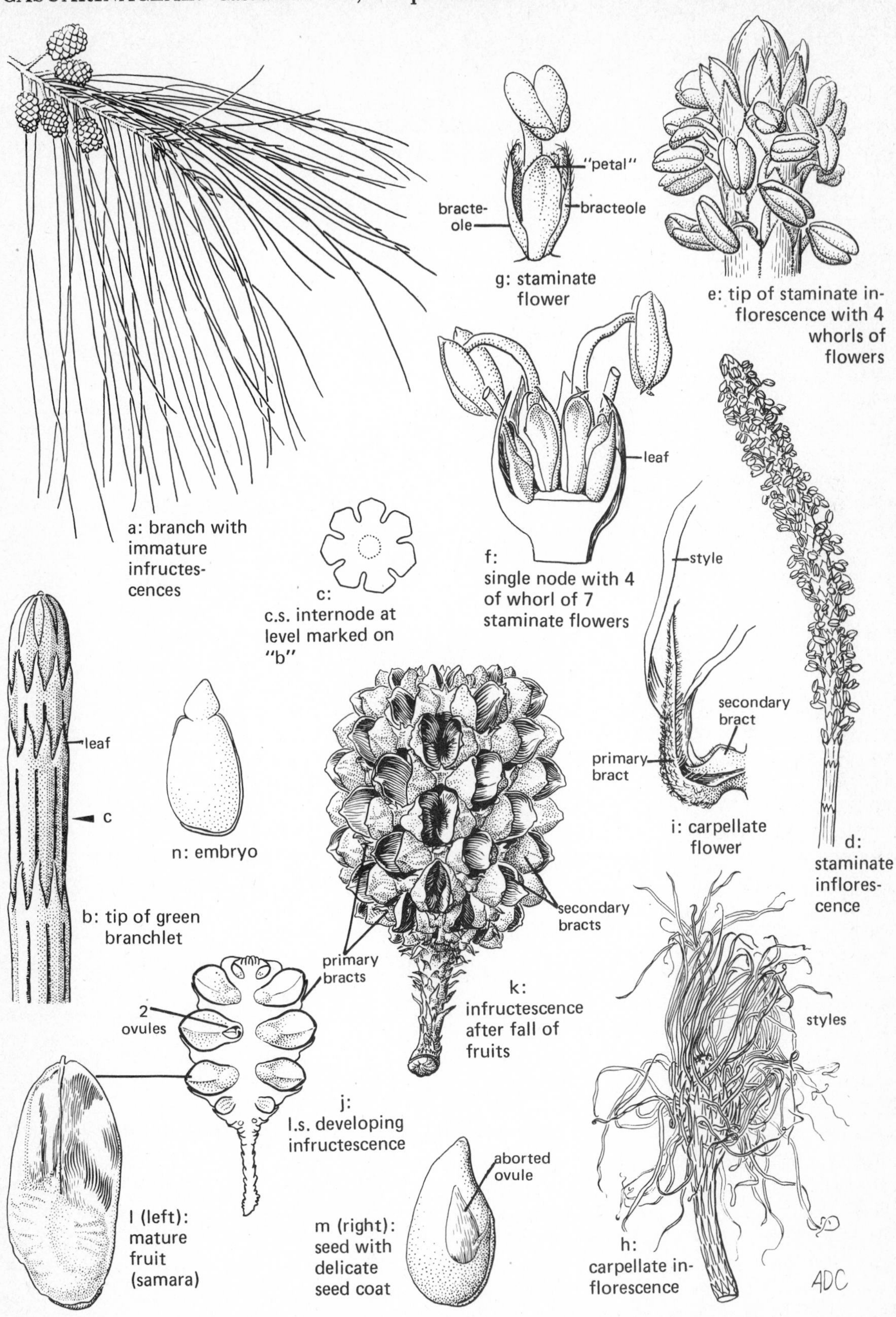

g: staminate flower

"petal"

bracte-ole

bracteole

e: tip of staminate in-florescence with 4 whorls of flowers

leaf

f: single node with 4 of whorl of 7 staminate flowers

a: branch with immature infructes-cences

c: c.s. internode at level marked on "b"

style

secondary bract

primary bract

i: carpellate flower

d: staminate inflores-cence

leaf

c

n: embryo

b: tip of green branchlet

secondary bracts

primary bracts

k: infructescence after fall of fruits

styles

2 ovules

j: l.s. developing infructescence

aborted ovule

l (left): mature fruit (samara)

m (right): seed with delicate seed coat

h: carpellate in-florescence

ADC

SAURURACEAE: Saururus. a-h, S. cernuus

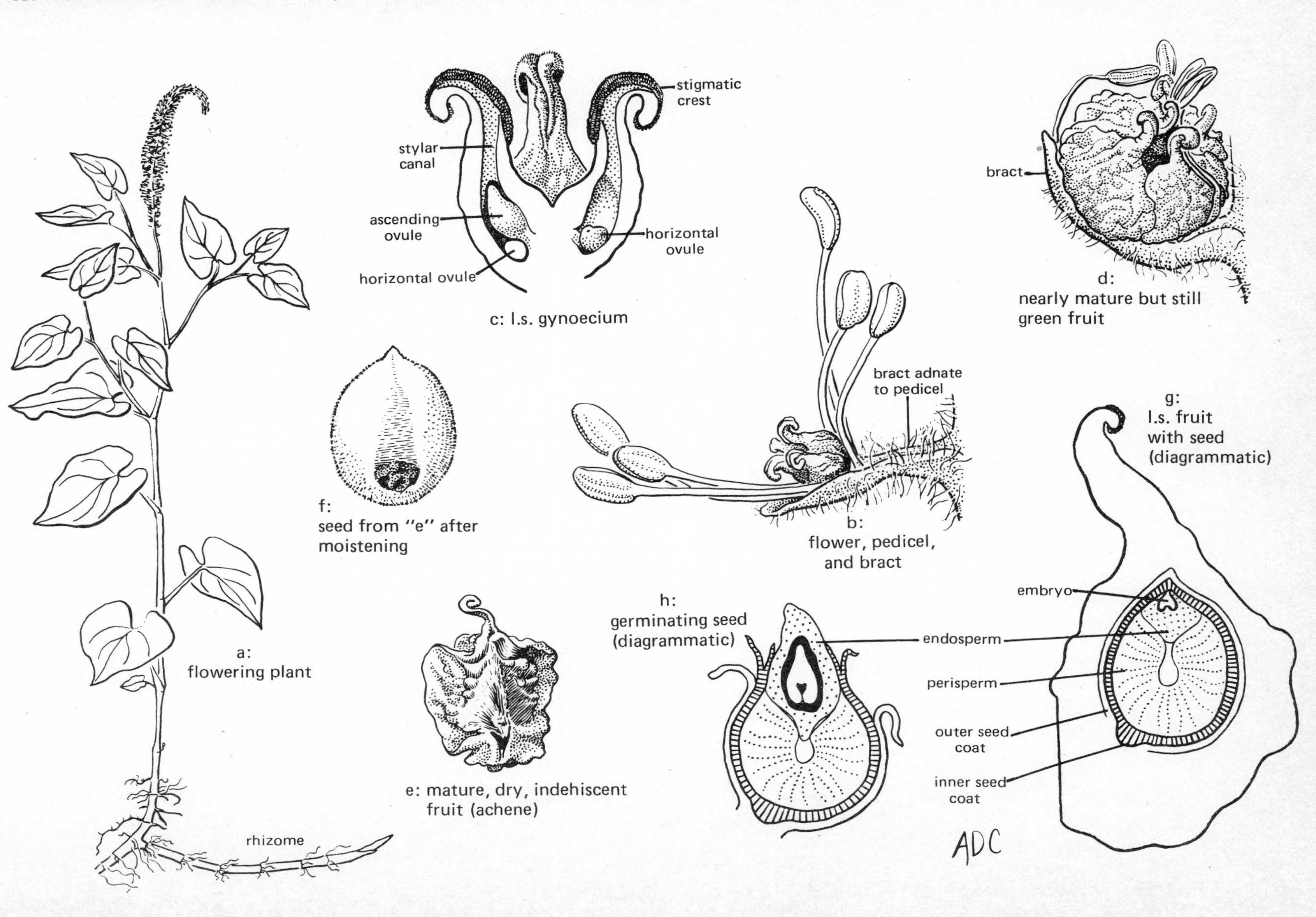

stigmatic crest

horizontal ovule

stylar canal

ascending ovule

horizontal ovule

c: l.s. gynoecium

bract adnate to pedicel

b: flower, pedicel, and bract

bract

d: nearly mature but still green fruit

embryo

endosperm

perisperm

outer seed coat

inner seed coat

g: l.s. fruit with seed (diagrammatic)

ADC

h: germinating seed (diagrammatic)

f: seed from "e" after moistening

e: mature, dry, indehiscent fruit (achene)

a: flowering plant

rhizome

2

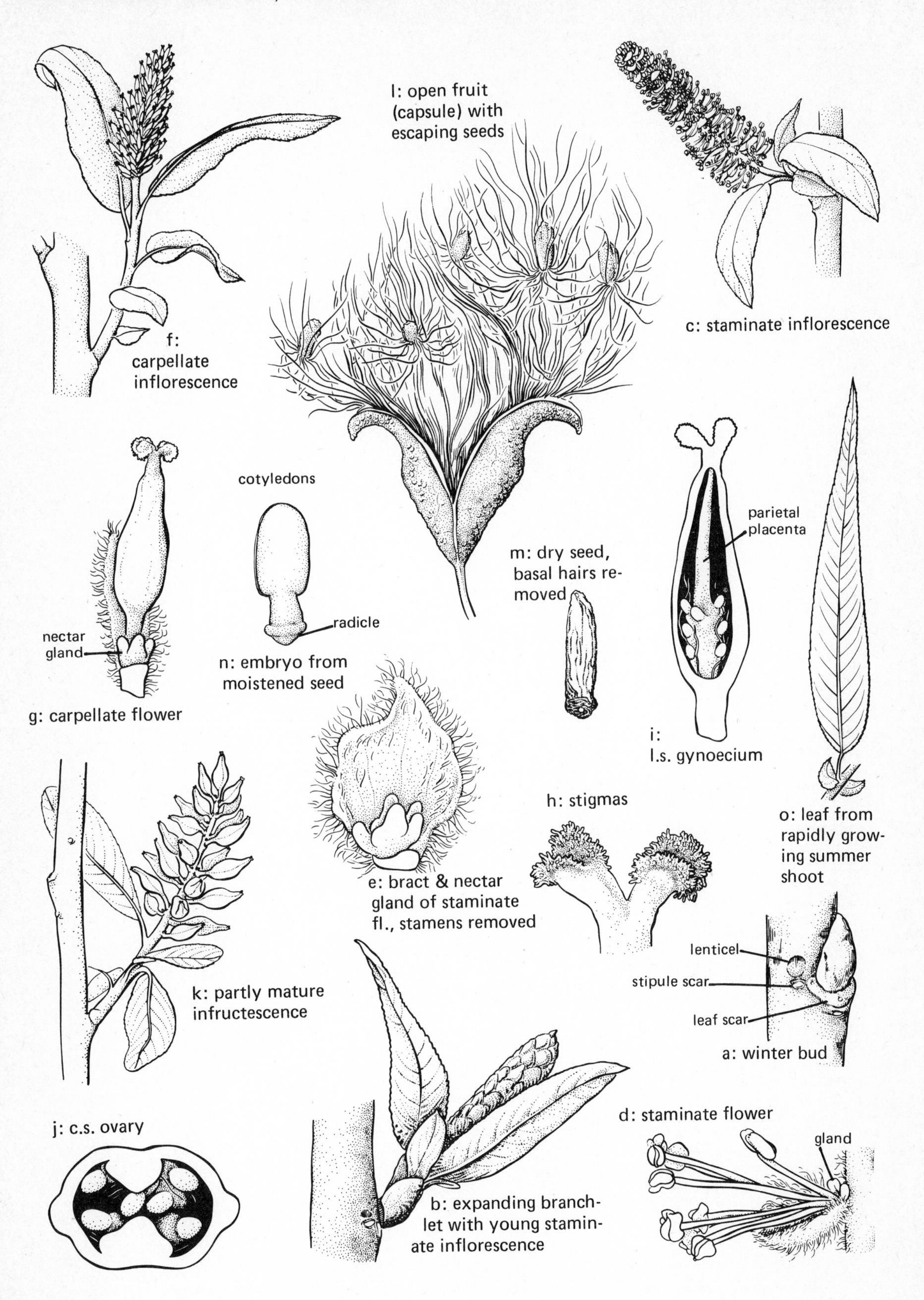

l: open fruit (capsule) with escaping seeds

f: carpellate inflorescence

c: staminate inflorescence

cotyledons

radicle

nectar gland

g: carpellate flower

n: embryo from moistened seed

m: dry seed, basal hairs removed

parietal placenta

i: l.s. gynoecium

o: leaf from rapidly growing summer shoot

h: stigmas

e: bract & nectar gland of staminate fl., stamens removed

k: partly mature infructescence

lenticel

stipule scar

leaf scar

a: winter bud

d: staminate flower

gland

j: c.s. ovary

b: expanding branchlet with young staminate inflorescence

SALICACEAE: Populus. a-j, P. deltoides; k, P. heterophylla

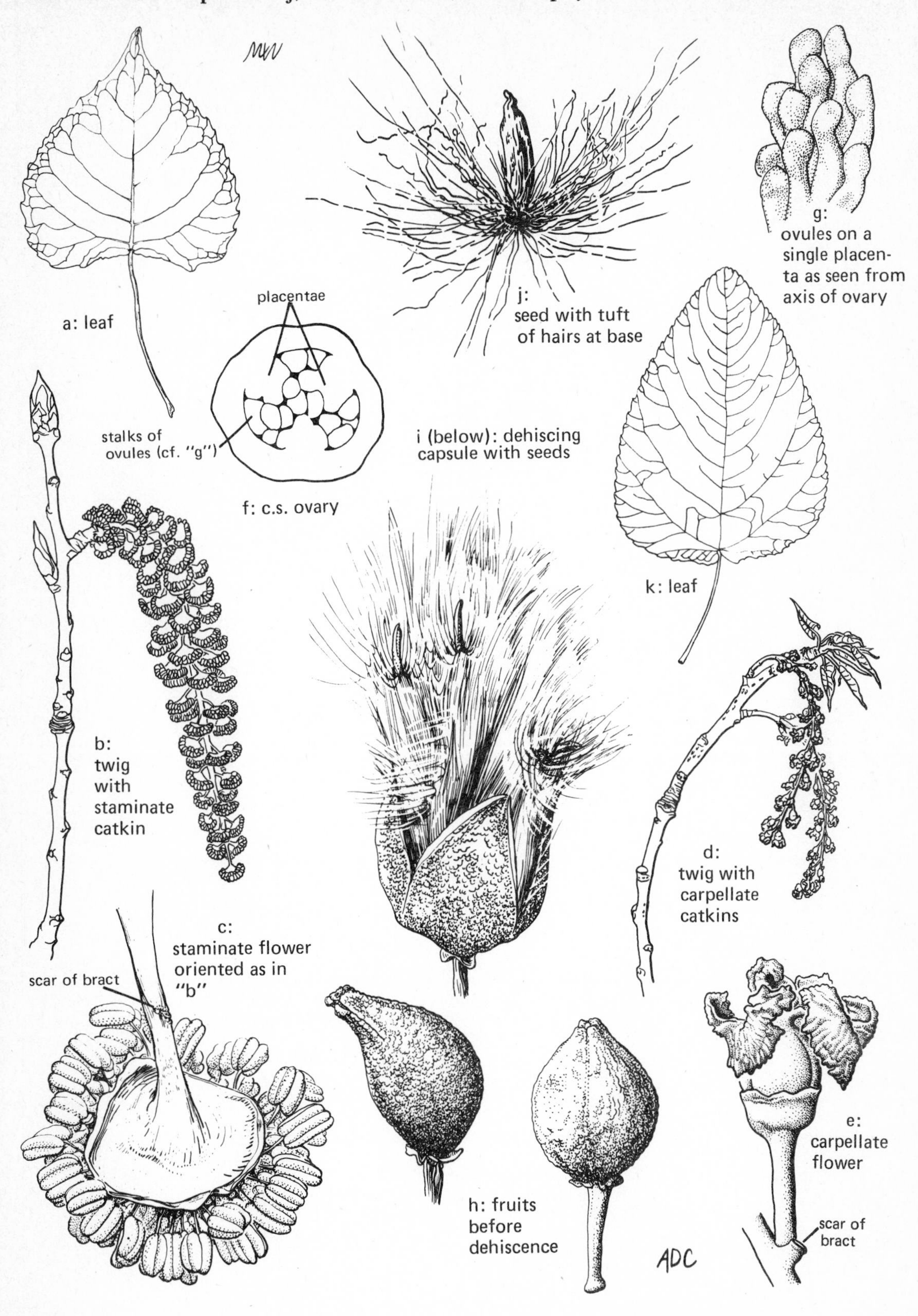

a: leaf

placentae

stalks of
ovules (cf. "g")

f: c.s. ovary

j:
seed with tuft
of hairs at base

i (below): dehiscing
capsule with seeds

g:
ovules on a
single placen-
ta as seen from
axis of ovary

k: leaf

b:
twig
with
staminate
catkin

d:
twig with
carpellate
catkins

scar of bract

c:
staminate flower
oriented as in
"b"

h: fruits
before
dehiscence

e:
carpellate
flower

scar of
bract

ADC

4

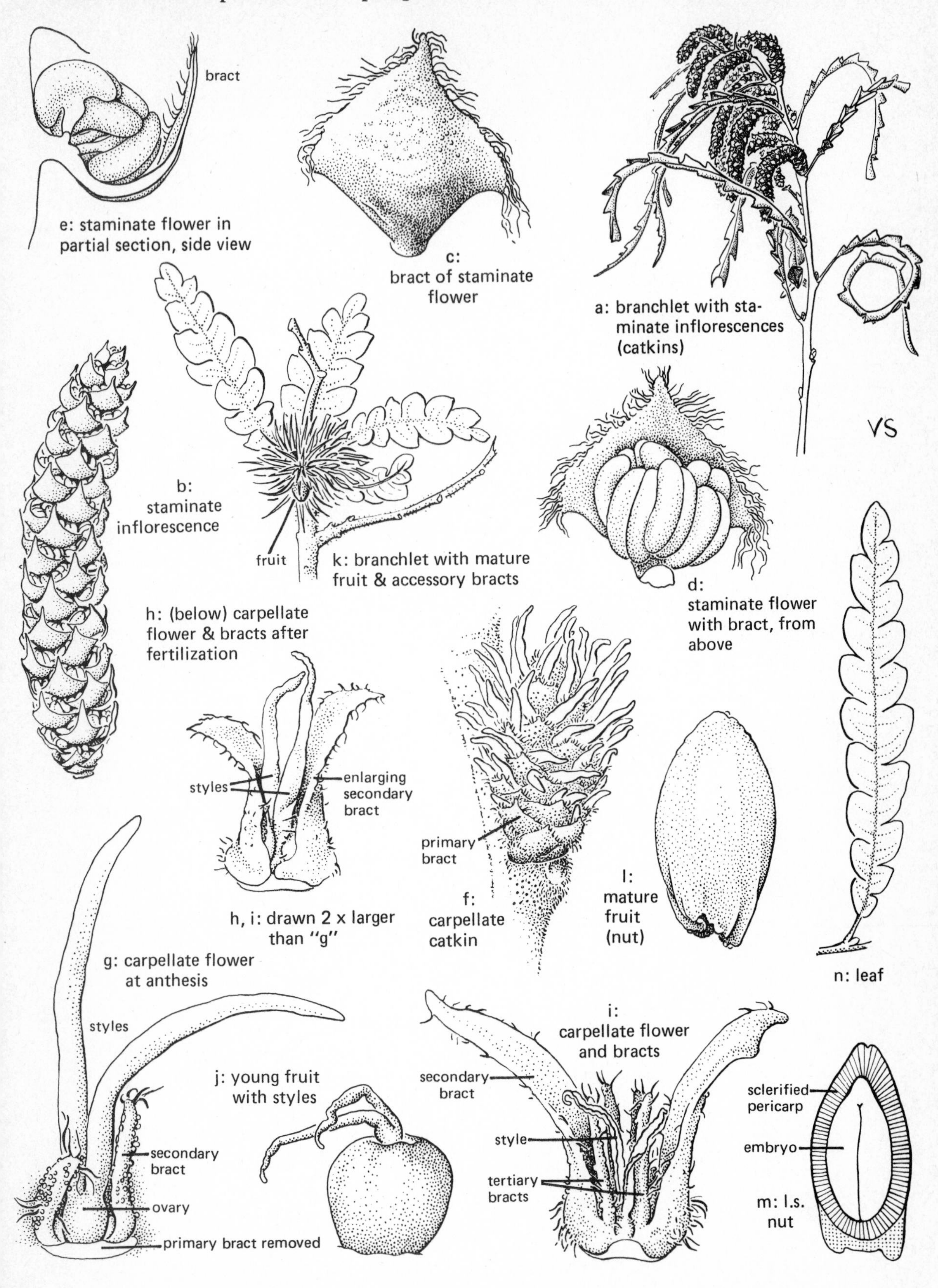

e: staminate flower in partial section, side view

c: bract of staminate flower

a: branchlet with staminate inflorescences (catkins)

b: staminate inflorescence

fruit

k: branchlet with mature fruit & accessory bracts

d: staminate flower with bract, from above

h: (below) carpellate flower & bracts after fertilization

styles

enlarging secondary bract

primary bract

f: carpellate catkin

l: mature fruit (nut)

n: leaf

h, i: drawn 2 x larger than "g"

g: carpellate flower at anthesis

styles

i: carpellate flower and bracts

j: young fruit with styles

secondary bract

secondary bract

style

tertiary bracts

sclerified pericarp

embryo

m: l.s. nut

ovary

primary bract removed

vs

5

JUGLANDACEAE: Carya.   a-m, C. ovata var. australis; n-s, C. laciniosa

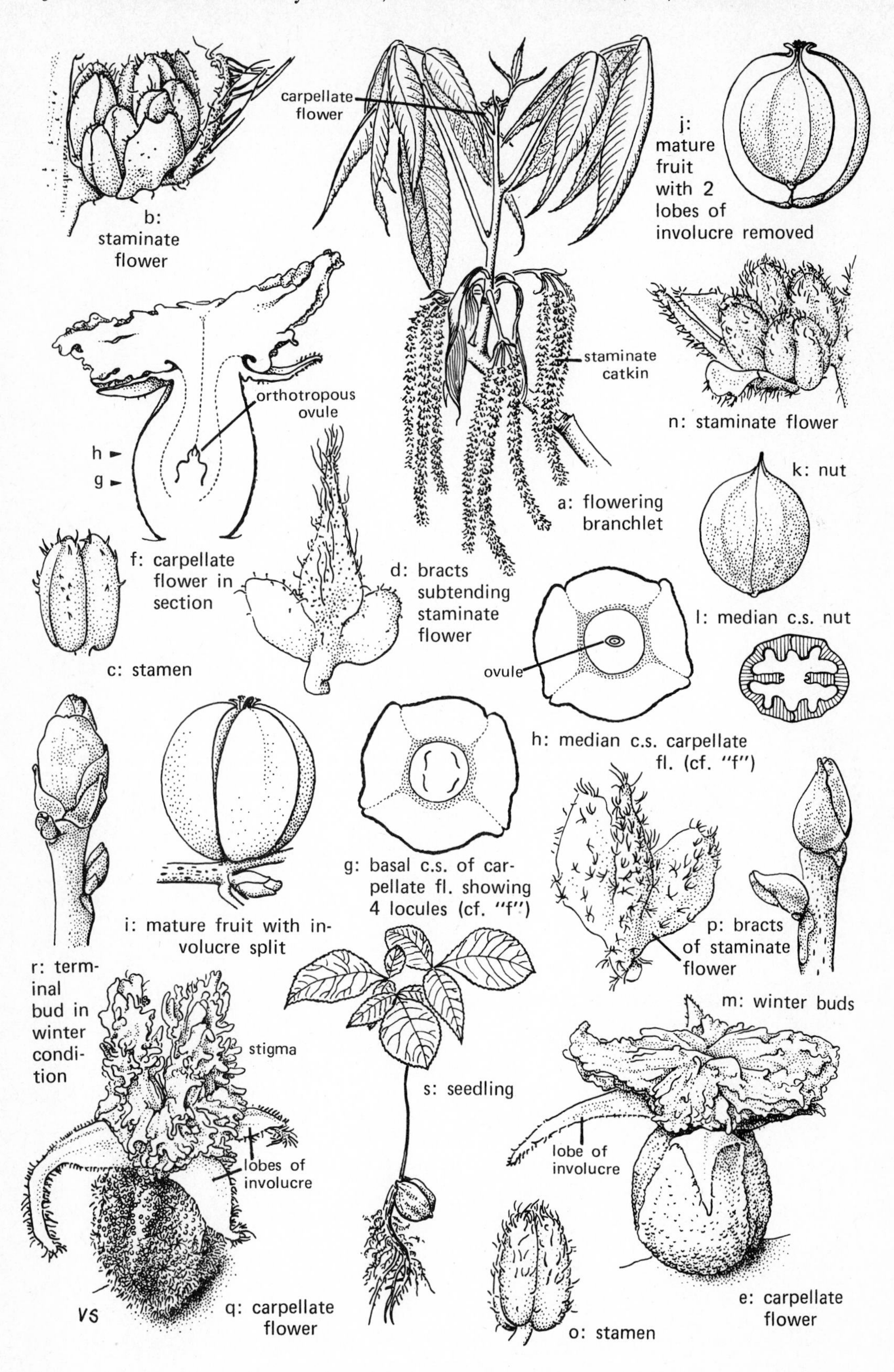

b: staminate flower

carpellate flower

j: mature fruit with 2 lobes of involucre removed

orthotropous ovule

h ▸
g ▸

staminate catkin

n: staminate flower

f: carpellate flower in section

c: stamen

d: bracts subtending staminate flower

a: flowering branchlet

k: nut

l: median c.s. nut

ovule

h: median c.s. carpellate fl. (cf. "f")

i: mature fruit with involucre split

g: basal c.s. of carpellate fl. showing 4 locules (cf. "f")

p: bracts of staminate flower

m: winter buds

r: terminal bud in winter condition

stigma

s: seedling

lobes of involucre

lobe of involucre

VS

q: carpellate flower

o: stamen

e: carpellate flower

6

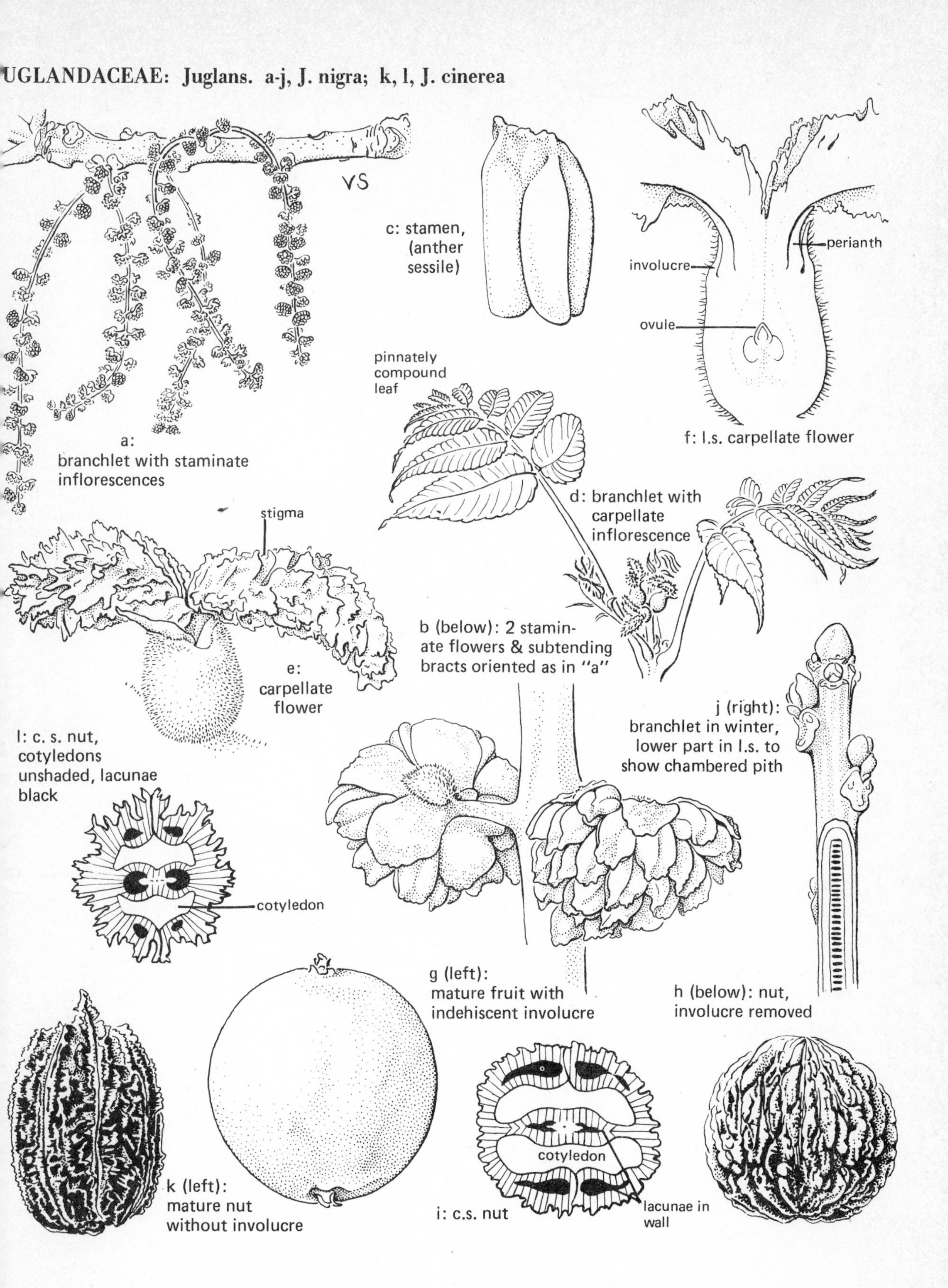

UGLANDACEAE: Juglans. a-j, J. nigra; k, l, J. cinerea

vs

c: stamen,
(anther
sessile)

perianth

involucre

ovule

f: l.s. carpellate flower

a:
branchlet with staminate
inflorescences

pinnately
compound
leaf

d: branchlet with
carpellate
inflorescence

stigma

b (below): 2 stamin-
ate flowers & subtending
bracts oriented as in "a"

e:
carpellate
flower

j (right):
branchlet in winter,
lower part in l.s. to
show chambered pith

l: c. s. nut,
cotyledons
unshaded, lacunae
black

cotyledon

g (left):
mature fruit with
indehiscent involucre

h (below): nut,
involucre removed

cotyledon

k (left):
mature nut
without involucre

i: c.s. nut

lacunae in
wall

BETULACEAE: Alnus. a-m, A. serrulata

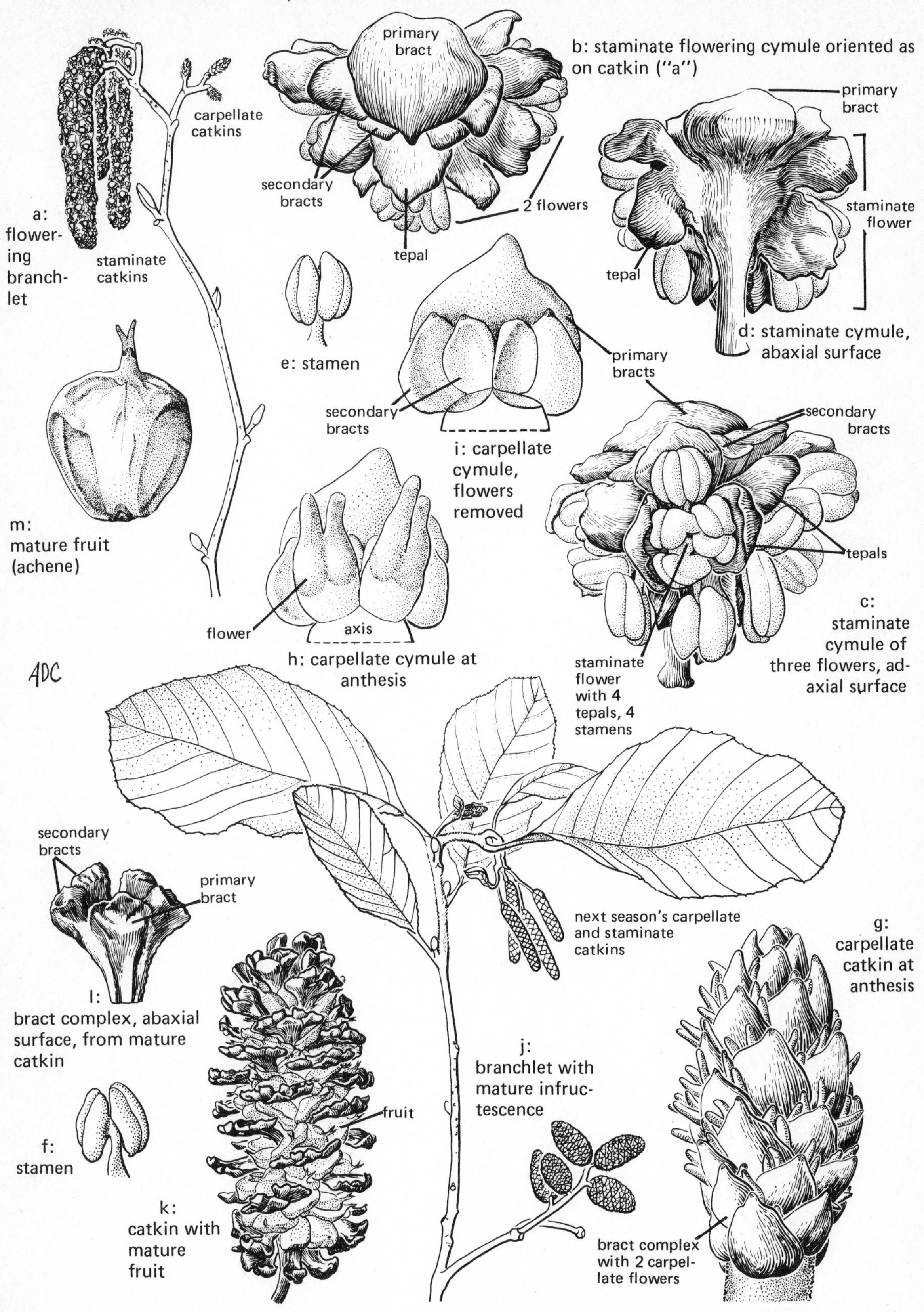

primary bract

b: staminate flowering cymule oriented as on catkin ("a")

primary bract

carpellate catkins

secondary bracts

2 flowers

tepal

a: flowering branchlet

staminate catkins

staminate flower

tepal

d: staminate cymule, abaxial surface

e: stamen

primary bracts

secondary bracts

secondary bracts

m: mature fruit (achene)

secondary bracts

i: carpellate cymule, flowers removed

tepals

c: staminate cymule of three flowers, adaxial surface

flower

axis

h: carpellate cymule at anthesis

staminate flower with 4 tepals, 4 stamens

ADC

secondary bracts

primary bract

next season's carpellate and staminate catkins

g: carpellate catkin at anthesis

l: bract complex, abaxial surface, from mature catkin

j: branchlet with mature infructescence

f: stamen

fruit

k: catkin with mature fruit

bract complex with 2 carpellate flowers

8

BETULACEAE: Betula. a-j, B. nigra; k, l, B. lenta

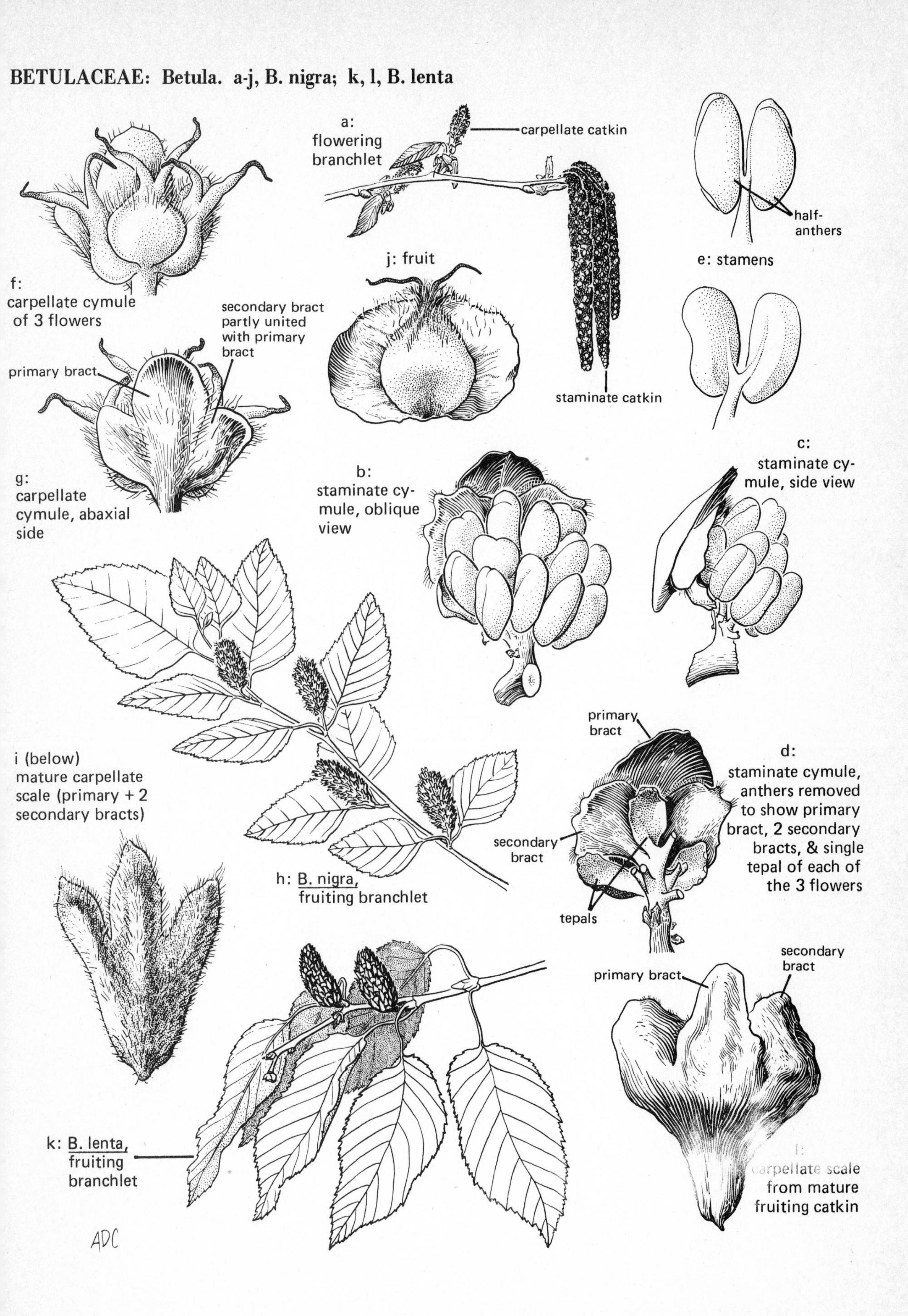

f: carpellate cymule of 3 flowers

a: flowering branchlet — carpellate catkin

j: fruit

staminate catkin

half-anthers

e: stamens

secondary bract partly united with primary bract

primary bract

g: carpellate cymule, abaxial side

b: staminate cymule, oblique view

c: staminate cymule, side view

i (below) mature carpellate scale (primary + 2 secondary bracts)

primary bract

secondary bract

d: staminate cymule, anthers removed to show primary bract, 2 secondary bracts, & single tepal of each of the 3 flowers

tepals

h: <u>B. nigra,</u> fruiting branchlet

secondary bract

primary bract

l: carpellate scale from mature fruiting catkin

k: <u>B. lenta,</u> fruiting branchlet

ADC

9

BETULACEAE: Ostrya & Carpinus. a-j, O. virginiana; k-p, C. caroliniana

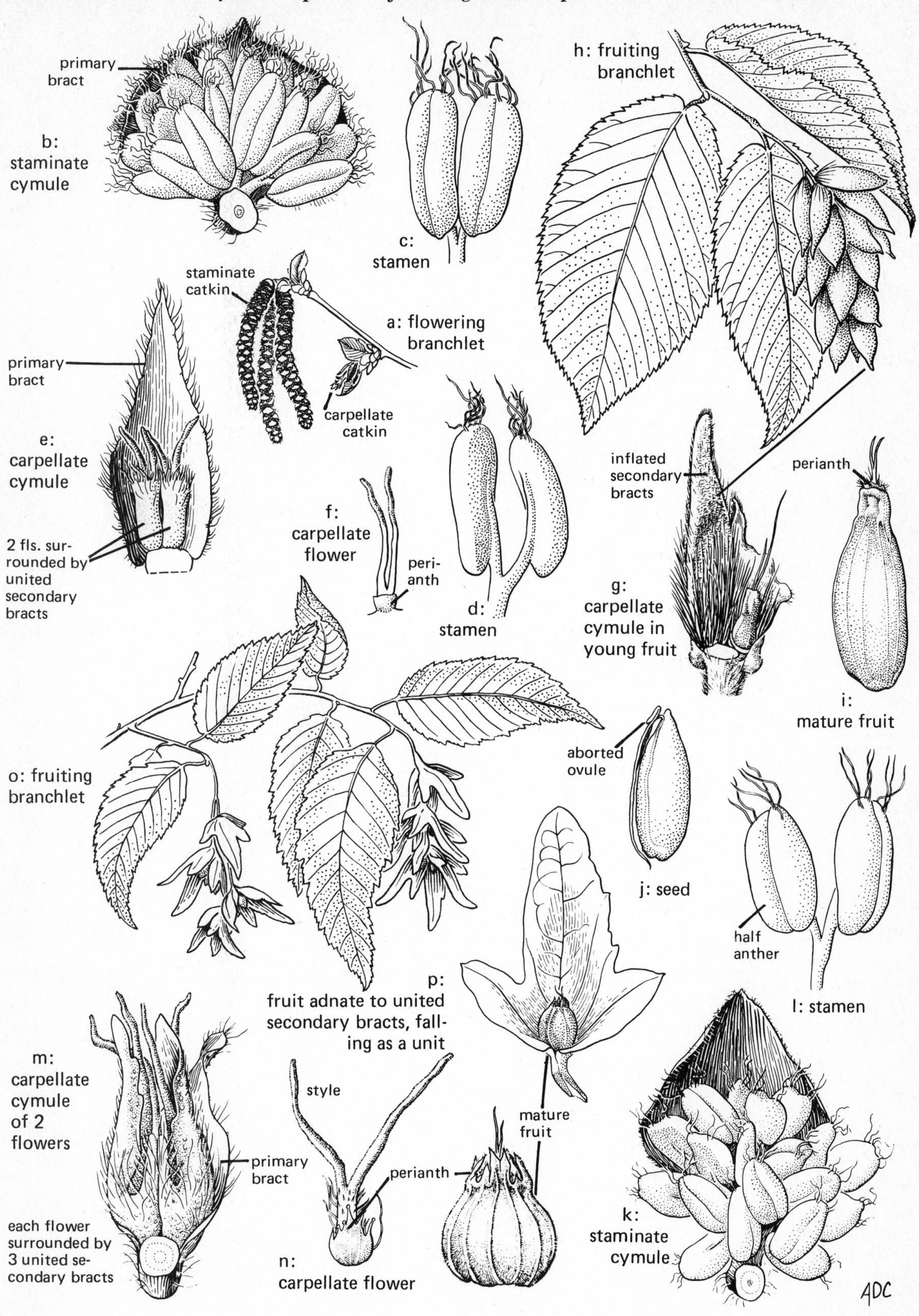

primary bract

b: staminate cymule

c: stamen

h: fruiting branchlet

staminate catkin

a: flowering branchlet

carpellate catkin

primary bract

e: carpellate cymule

2 fls. surrounded by united secondary bracts

f: carpellate flower

peri- anth

d: stamen

inflated secondary bracts

perianth

g: carpellate cymule in young fruit

i: mature fruit

o: fruiting branchlet

aborted ovule

j: seed

half anther

l: stamen

p: fruit adnate to united secondary bracts, fall- ing as a unit

m: carpellate cymule of 2 flowers

style

primary bract

perianth

mature fruit

k: staminate cymule

each flower surrounded by 3 united se- condary bracts

n: carpellate flower

ADC

10

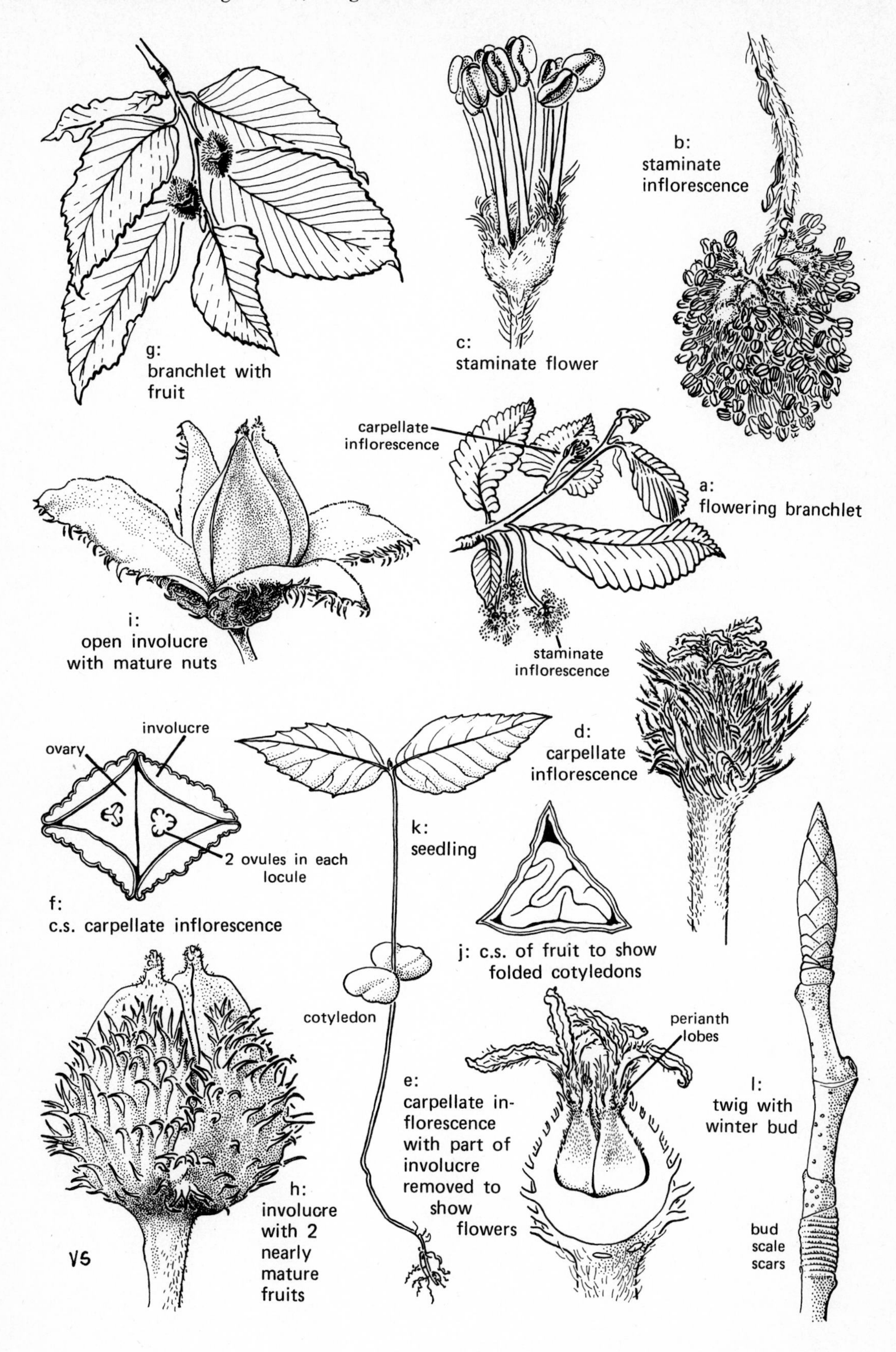

b:
staminate
inflorescence

g:
branchlet with
fruit

c:
staminate flower

carpellate
inflorescence

a:
flowering branchlet

i:
open involucre
with mature nuts

staminate
inflorescence

d:
carpellate
inflorescence

ovary

involucre

2 ovules in each
locule

f:
c.s. carpellate inflorescence

k:
seedling

j: c.s. of fruit to show
folded cotyledons

perianth
lobes

l:
twig with
winter bud

cotyledon

e:
carpellate in-
florescence
with part of
involucre
removed to
show
flowers

h:
involucre
with 2
nearly
mature
fruits

√5

bud
scale
scars

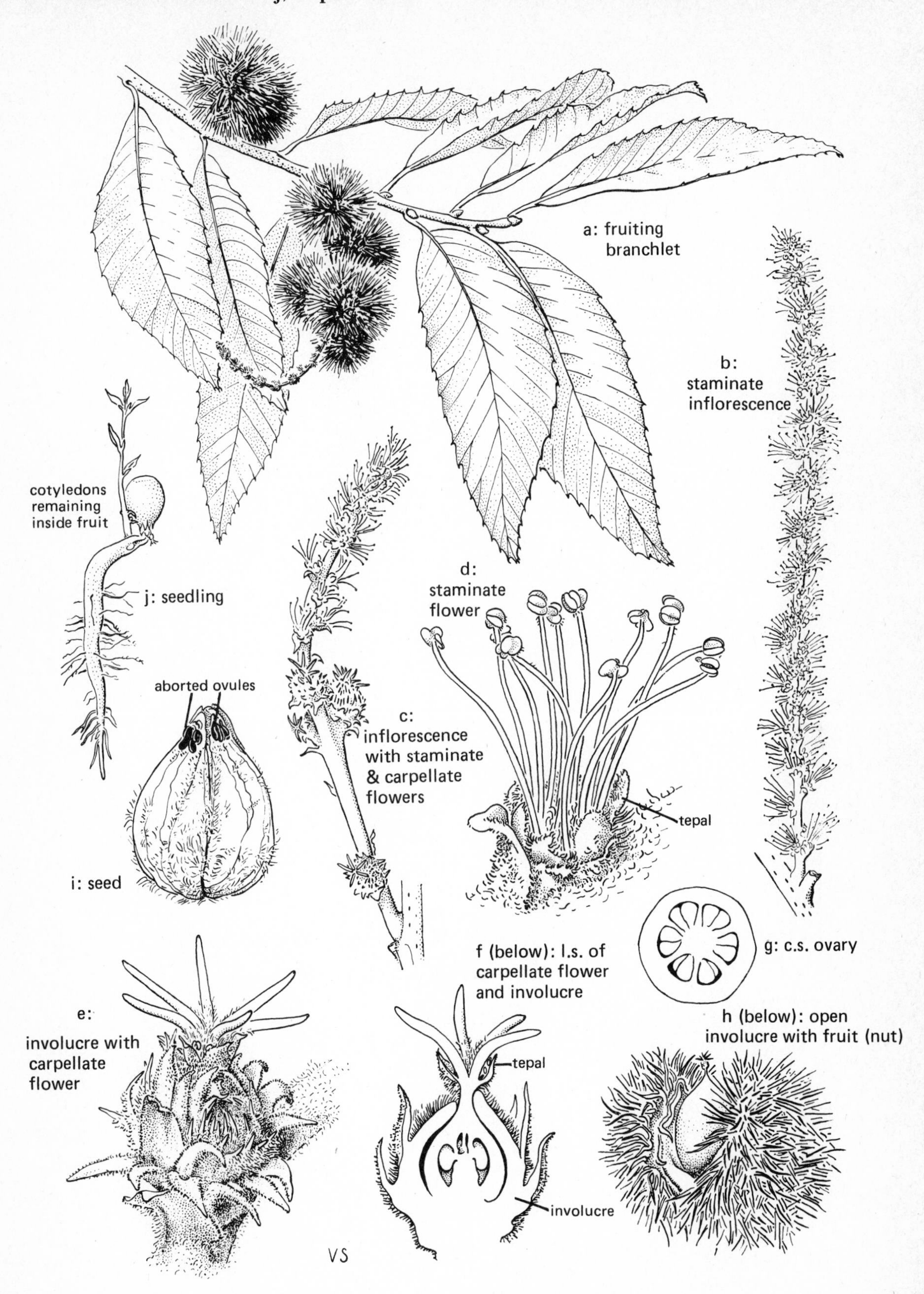

a: fruiting
branchlet

b:
staminate
inflorescence

cotyledons
remaining
inside fruit

j: seedling

d:
staminate
flower

aborted ovules

c:
inflorescence
with staminate
& carpellate
flowers

tepal

i: seed

f (below): l.s. of
carpellate flower
and involucre

g: c.s. ovary

e:
involucre with
carpellate
flower

tepal

h (below): open
involucre with fruit (nut)

involucre

VS

12

FAGACEAE: Quercus subgenus Quercus. a-d, Q. prinus; e-g, Q. alba; h, i, Q. macrocarpa; j, k, Q. muhlenbergii; l, m, Q. virginiana

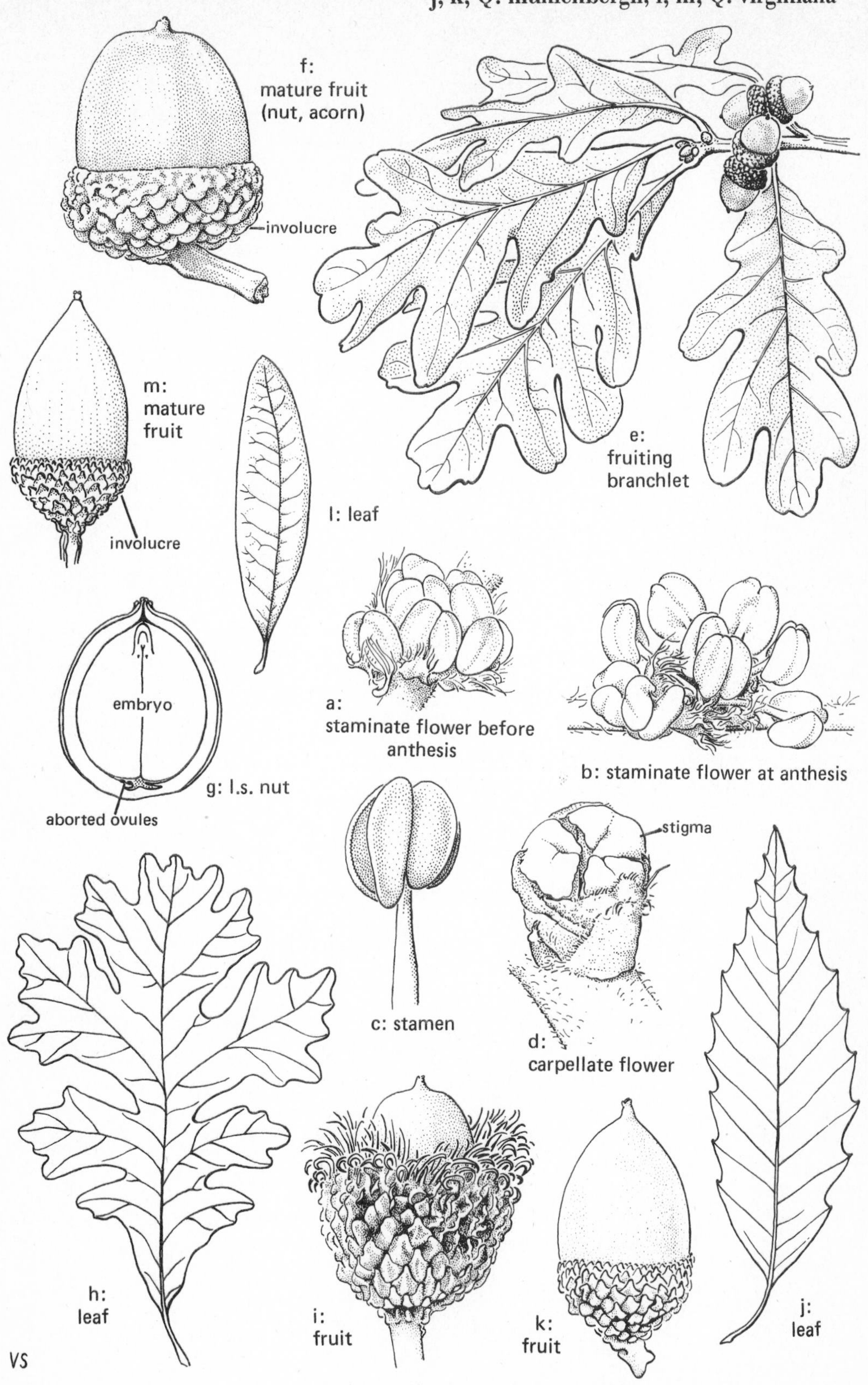

f: mature fruit (nut, acorn)

involucre

m: mature fruit

involucre

l: leaf

e: fruiting branchlet

embryo

aborted ovules

g: l.s. nut

a: staminate flower before anthesis

b: staminate flower at anthesis

c: stamen

stigma

d: carpellate flower

h: leaf

i: fruit

k: fruit

j: leaf

VS

FAGACEAE: Quercus subgenus Erythrobalanus. a-e, Q. ilicifolia; f-h, Q. rubra; i, j, Q. imbricata; k, l, Q. nigra

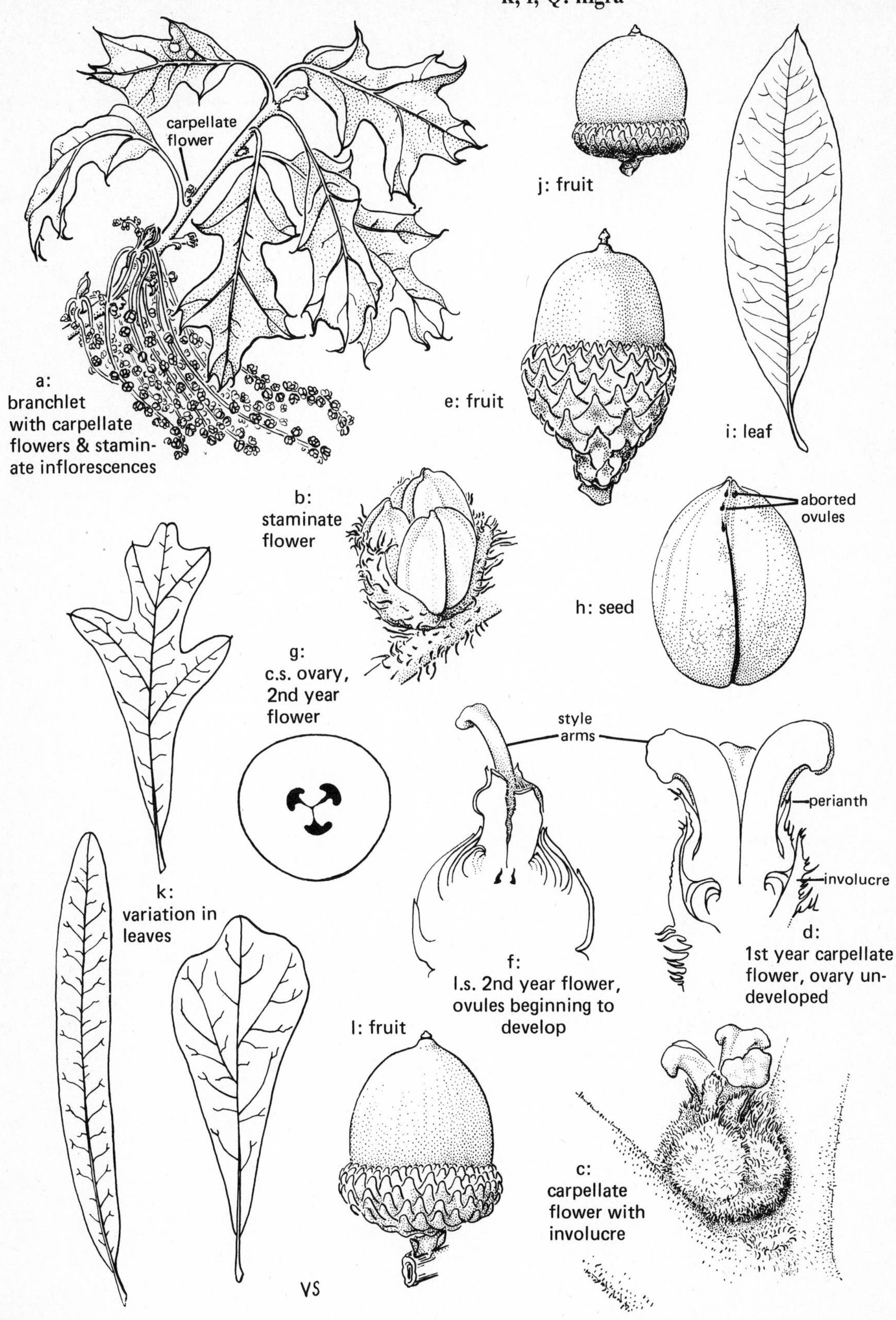

carpellate flower

a: branchlet with carpellate flowers & staminate inflorescences

b: staminate flower

g: c.s. ovary, 2nd year flower

k: variation in leaves

e: fruit

j: fruit

i: leaf

h: seed

aborted ovules

style arms

perianth

involucre

d: 1st year carpellate flower, ovary un-developed

f: l.s. 2nd year flower, ovules beginning to develop

l: fruit

c: carpellate flower with involucre

VS

14

ULMACEAE: Ulmus. a-h, U. americana; i-k, U. rubra; l, U. crassifolia; m, U. alata

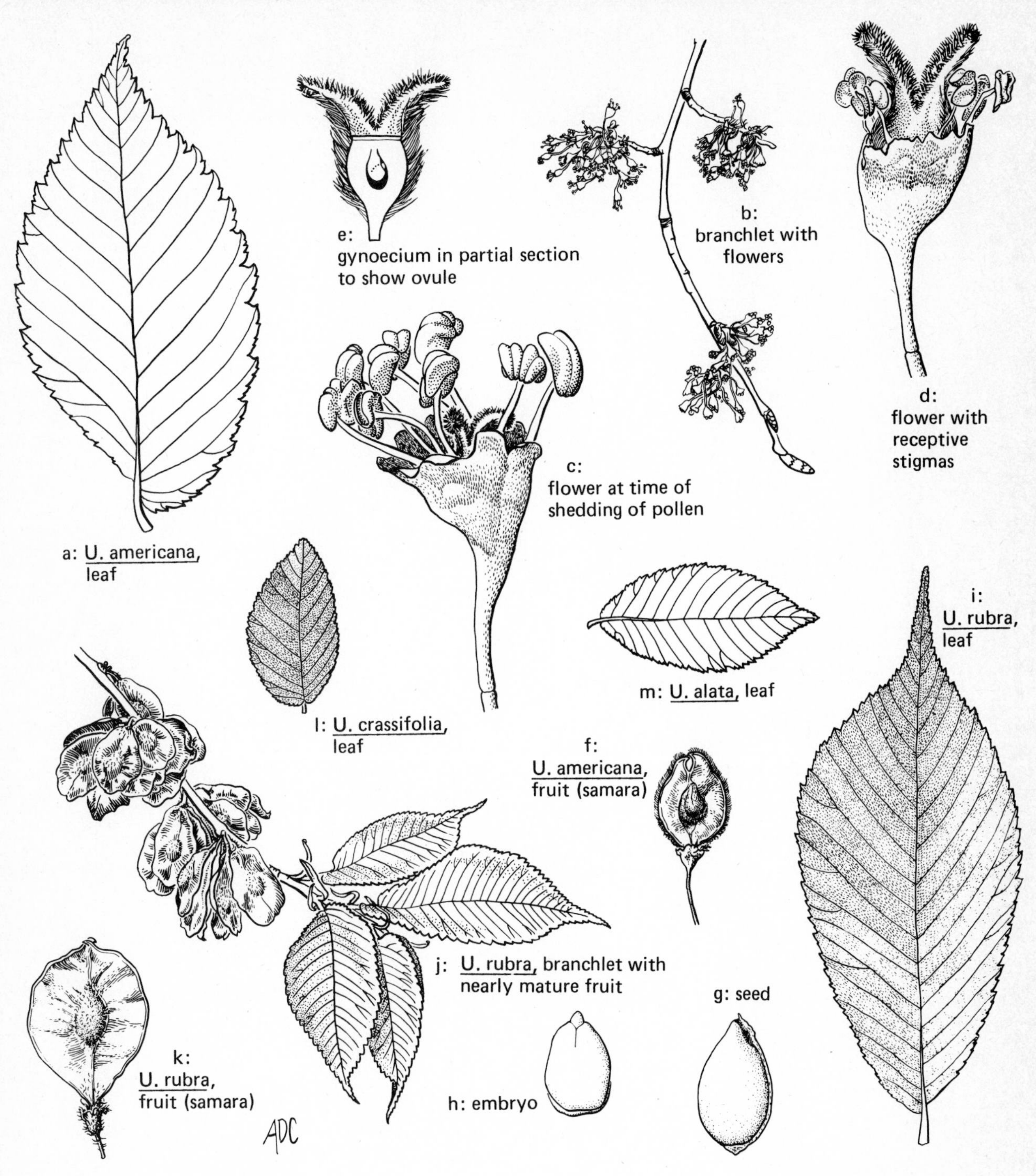

e:
gynoecium in partial section
to show ovule

b:
branchlet with
flowers

d:
flower with
receptive
stigmas

c:
flower at time of
shedding of pollen

a: U. americana,
leaf

l: U. crassifolia,
leaf

m: U. alata, leaf

i:
U. rubra,
leaf

f:
U. americana,
fruit (samara)

j: U. rubra, branchlet with
nearly mature fruit

g: seed

k:
U. rubra,
fruit (samara)

h: embryo

ADC

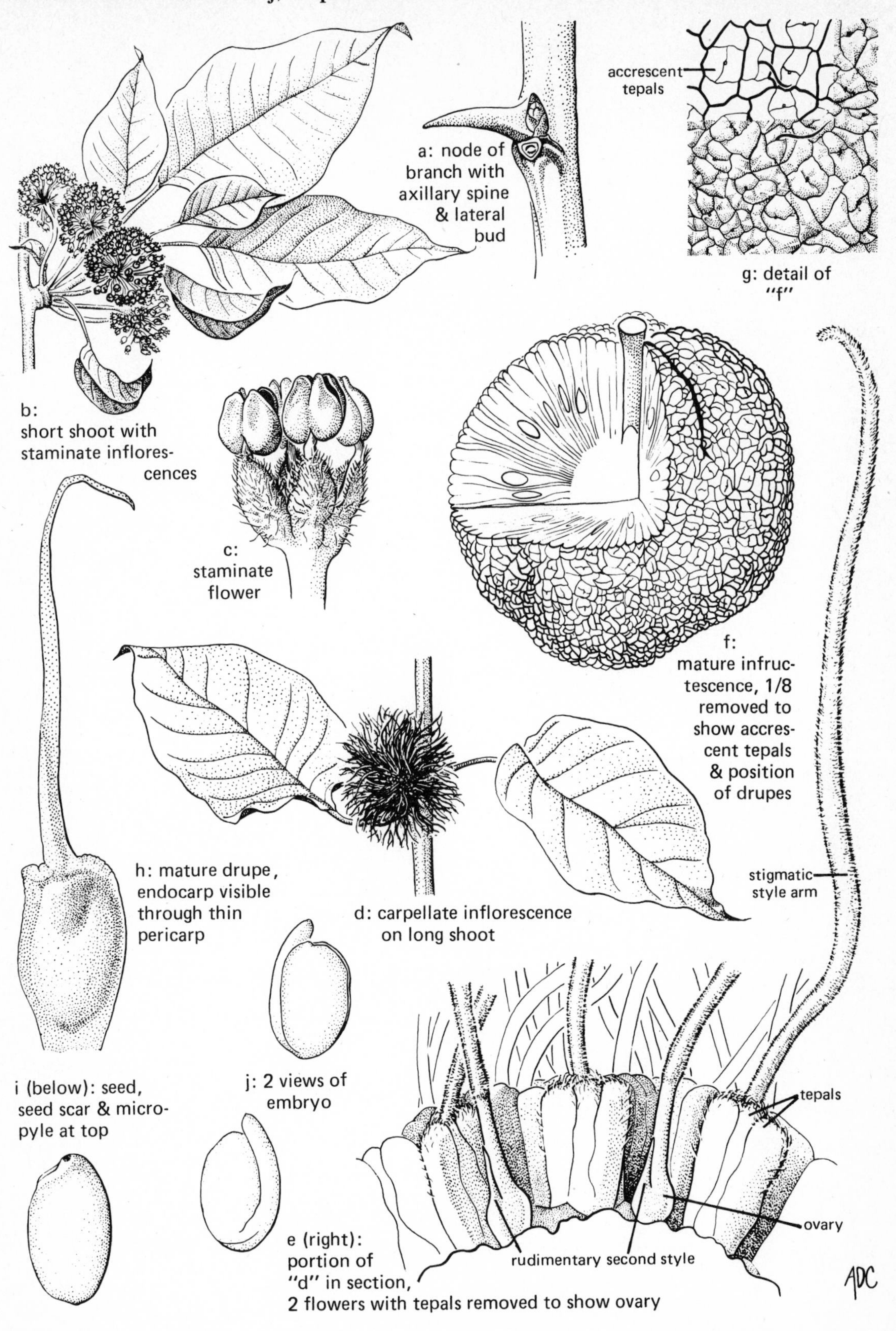

MORACEAE: Maclura. a-j, M. pomifera

a: node of branch with axillary spine & lateral bud

accrescent tepals

g: detail of "f"

b: short shoot with staminate inflorescences

c: staminate flower

f: mature infructescence, 1/8 removed to show accrescent tepals & position of drupes

stigmatic style arm

h: mature drupe, endocarp visible through thin pericarp

d: carpellate inflorescence on long shoot

i (below): seed, seed scar & micropyle at top

j: 2 views of embryo

tepals

ovary

rudimentary second style

e (right): portion of "d" in section, 2 flowers with tepals removed to show ovary

ADC

URTICACEAE: Urtica. a-k, U. chamaedryoides; l, U. dioica

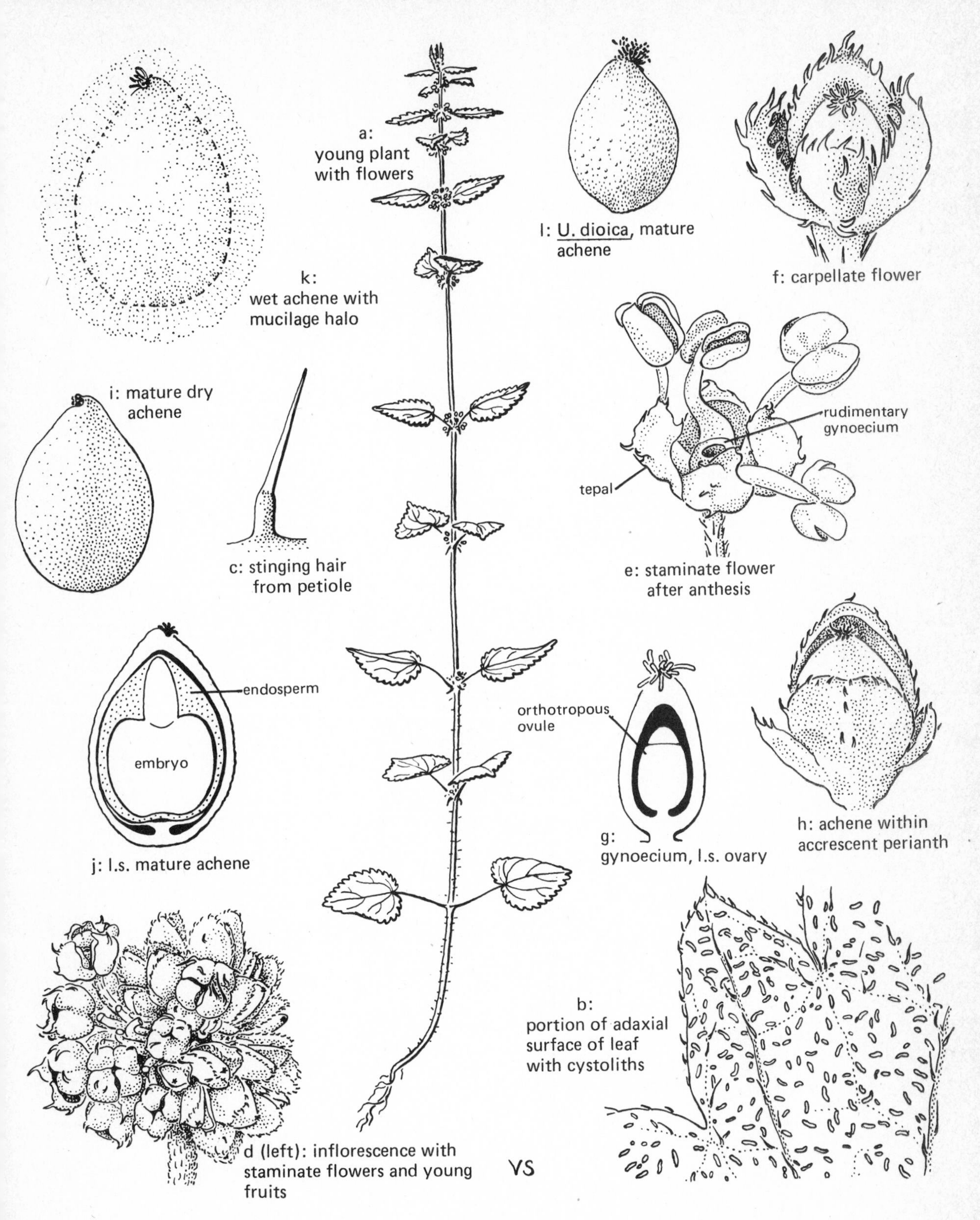

a:
young plant
with flowers

l: U. dioica, mature
achene

k:
wet achene with
mucilage halo

f: carpellate flower

i: mature dry
achene

rudimentary
gynoecium

tepal

c: stinging hair
from petiole

e: staminate flower
after anthesis

endosperm

embryo

orthotropous
ovule

j: l.s. mature achene

g:
gynoecium, l.s. ovary

h: achene within
accrescent perianth

b:
portion of adaxial
surface of leaf
with cystoliths

d (left): inflorescence with
staminate flowers and young
fruits

VS

17

URTICACEAE: Pilea. a-f, P. microphylla; g-n, P. pumila; o, P. fontana

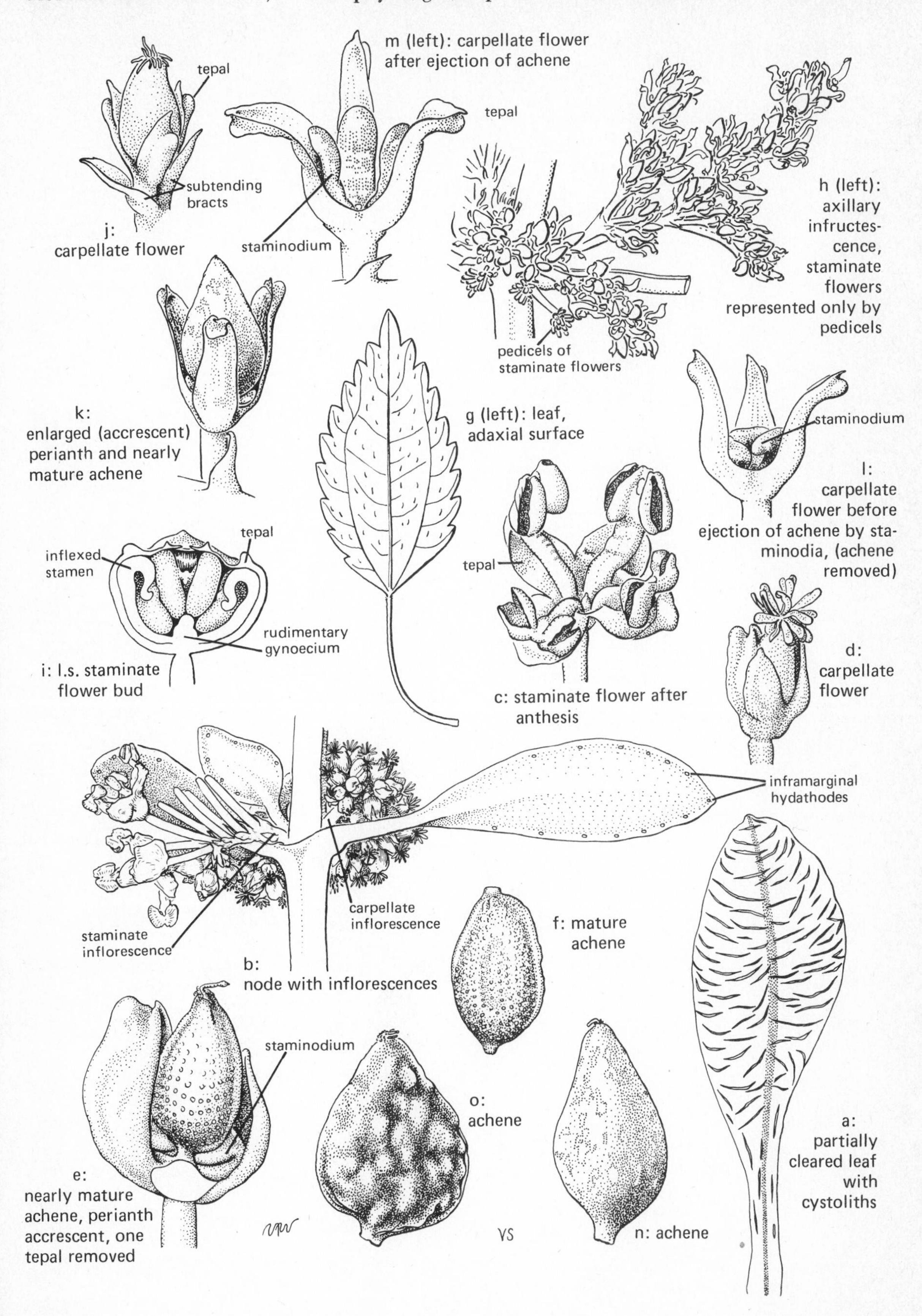

j: carpellate flower

tepal

subtending bracts

m (left): carpellate flower after ejection of achene

tepal

staminodium

h (left): axillary infructes-cence, staminate flowers represented only by pedicels

pedicels of staminate flowers

k: enlarged (accrescent) perianth and nearly mature achene

g (left): leaf, adaxial surface

staminodium

l: carpellate flower before ejection of achene by sta-minodia, (achene removed)

inflexed stamen

tepal

tepal

d: carpellate flower

rudimentary gynoecium

i: l.s. staminate flower bud

c: staminate flower after anthesis

inframarginal hydathodes

staminate inflorescence

carpellate inflorescence

f: mature achene

b: node with inflorescences

staminodium

o: achene

a: partially cleared leaf with cystoliths

e: nearly mature achene, perianth accrescent, one tepal removed

n: achene

18

CANNABACEAE: Cannabis. a-k, C. sativa

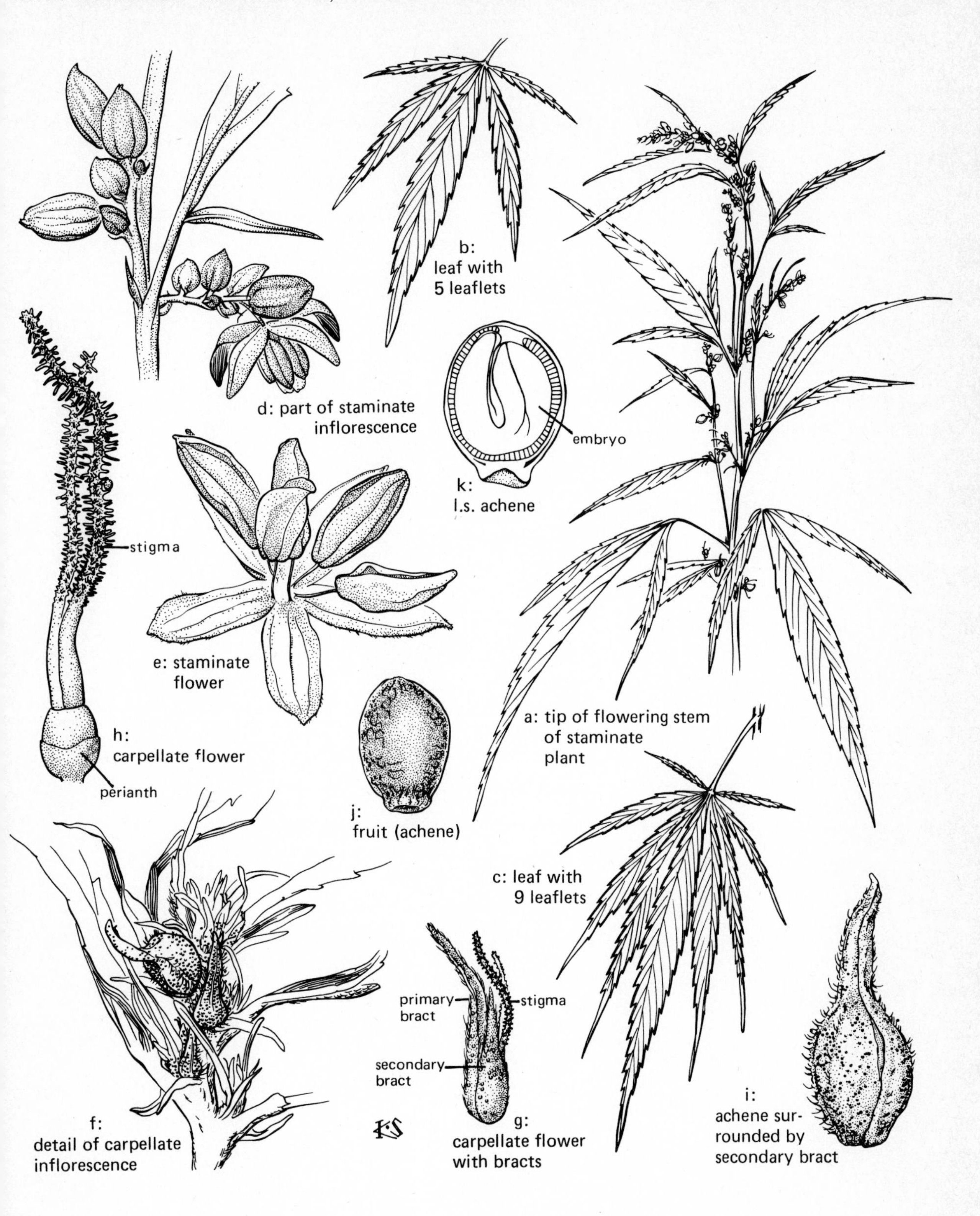

b: leaf with 5 leaflets

d: part of staminate inflorescence

k: l.s. achene

embryo

stigma

e: staminate flower

h: carpellate flower

perianth

j: fruit (achene)

a: tip of flowering stem of staminate plant

c: leaf with 9 leaflets

primary bract

stigma

secondary bract

f: detail of carpellate inflorescence

g: carpellate flower with bracts

i: achene surrounded by secondary bract

ARISTOLOCHIACEAE: Asarum. a-g, A. canadense; h-k, A. arifolium; l, A. ruthii; m, n, A. shuttleworthii; o, p, A. lewisii

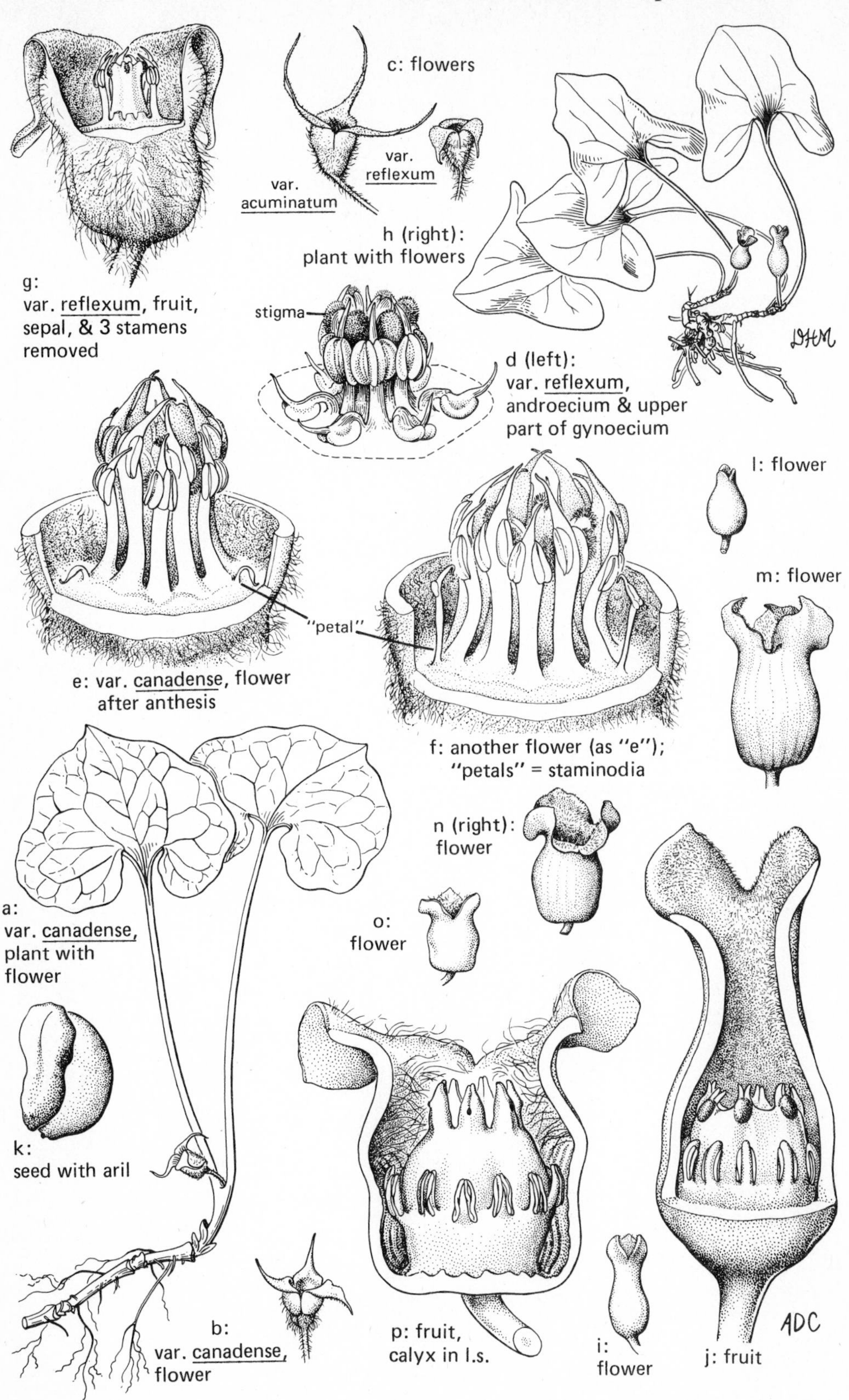

g: var. reflexum, fruit, sepal, & 3 stamens removed

c: flowers

var. reflexum

var. acuminatum

h (right): plant with flowers

stigma

d (left): var. reflexum, androecium & upper part of gynoecium

l: flower

m: flower

"petal"

e: var. canadense, flower after anthesis

f: another flower (as "e"); "petals" = staminodia

n (right): flower

a: var. canadense, plant with flower

o: flower

k: seed with aril

b: var. canadense, flower

p: fruit, calyx in l.s.

i: flower

j: fruit

20

ARISTOLOCHIACEAE: Aristolochia. a-i, A. tomentosa

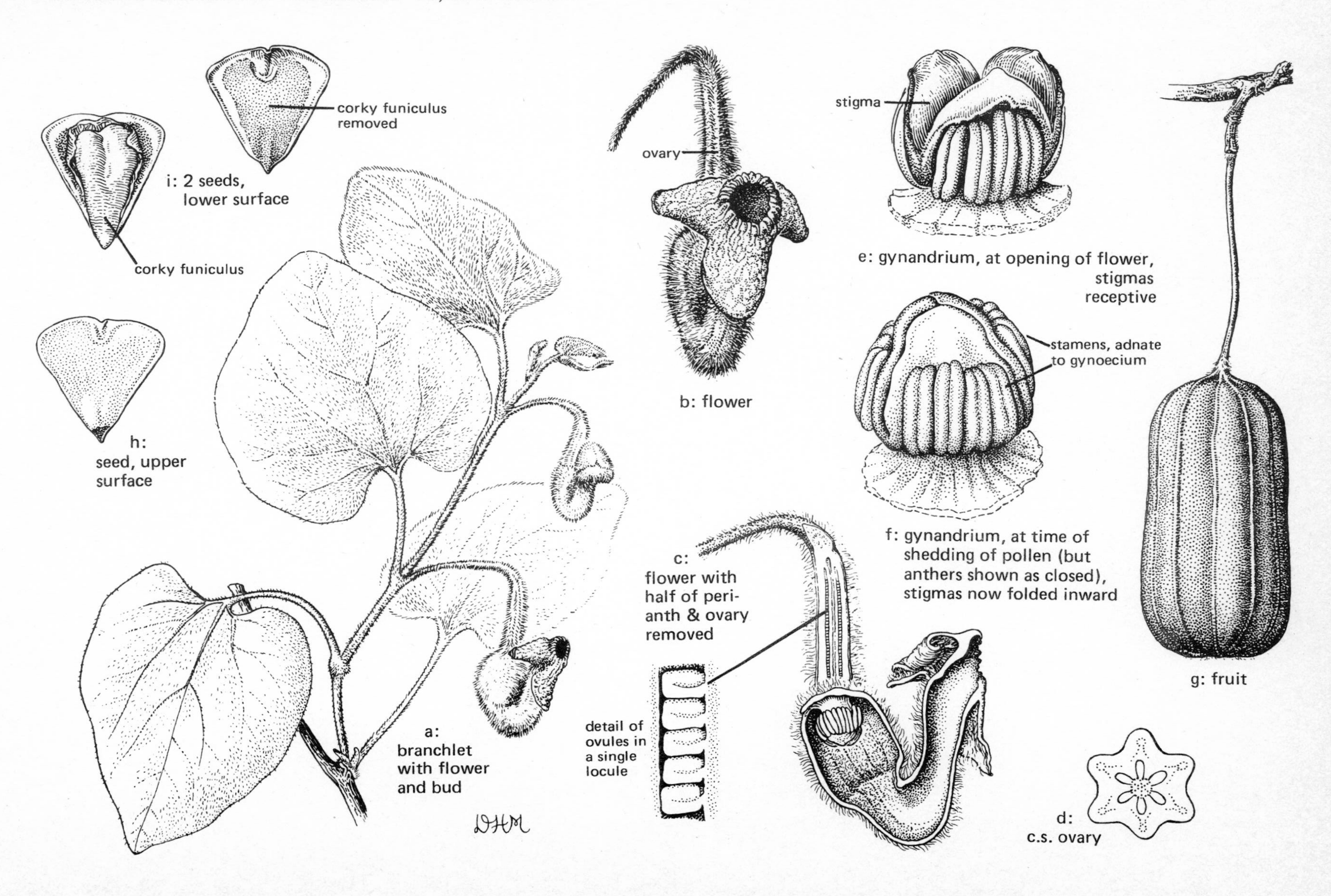

g: fruit

d: c.s. ovary

stigma

e: gynandrium, at opening of flower, stigmas receptive

stamens, adnate to gynoecium

f: gynandrium, at time of shedding of pollen (but anthers shown as closed), stigmas now folded inward

ovary

b: flower

c: flower with half of peri- anth & ovary removed

detail of ovules in a single locule

a: branchlet with flower and bud

corky funiculus removed

i: 2 seeds, lower surface

corky funiculus

h: seed, upper surface

21

POLYGONACEAE: Polygonum sections Tiniaria (a-j) & Echinocaulon (k,l). a-f, P. scandens; g-j, P. cuspidatum; k,l, P. sagittatum

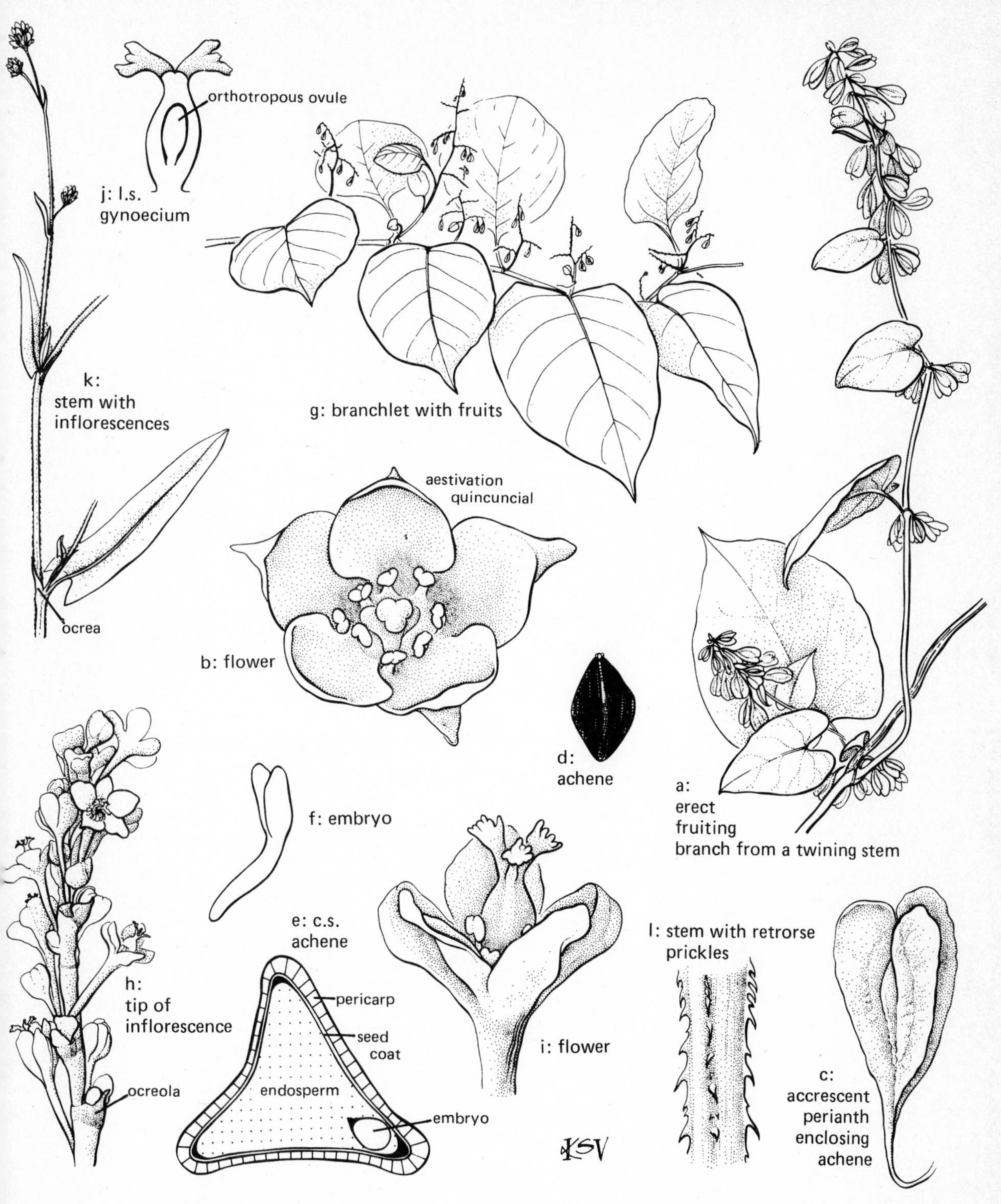

j: l.s. gynoecium

orthotropous ovule

k: stem with inflorescences

ocrea

g: branchlet with fruits

aestivation quincuncial

b: flower

d: achene

f: embryo

a: erect fruiting branch from a twining stem

e: c.s. achene

pericarp

seed coat

endosperm

embryo

h: tip of inflorescence

ocreola

i: flower

l: stem with retrorse prickles

c: accrescent perianth enclosing achene

KSV

22

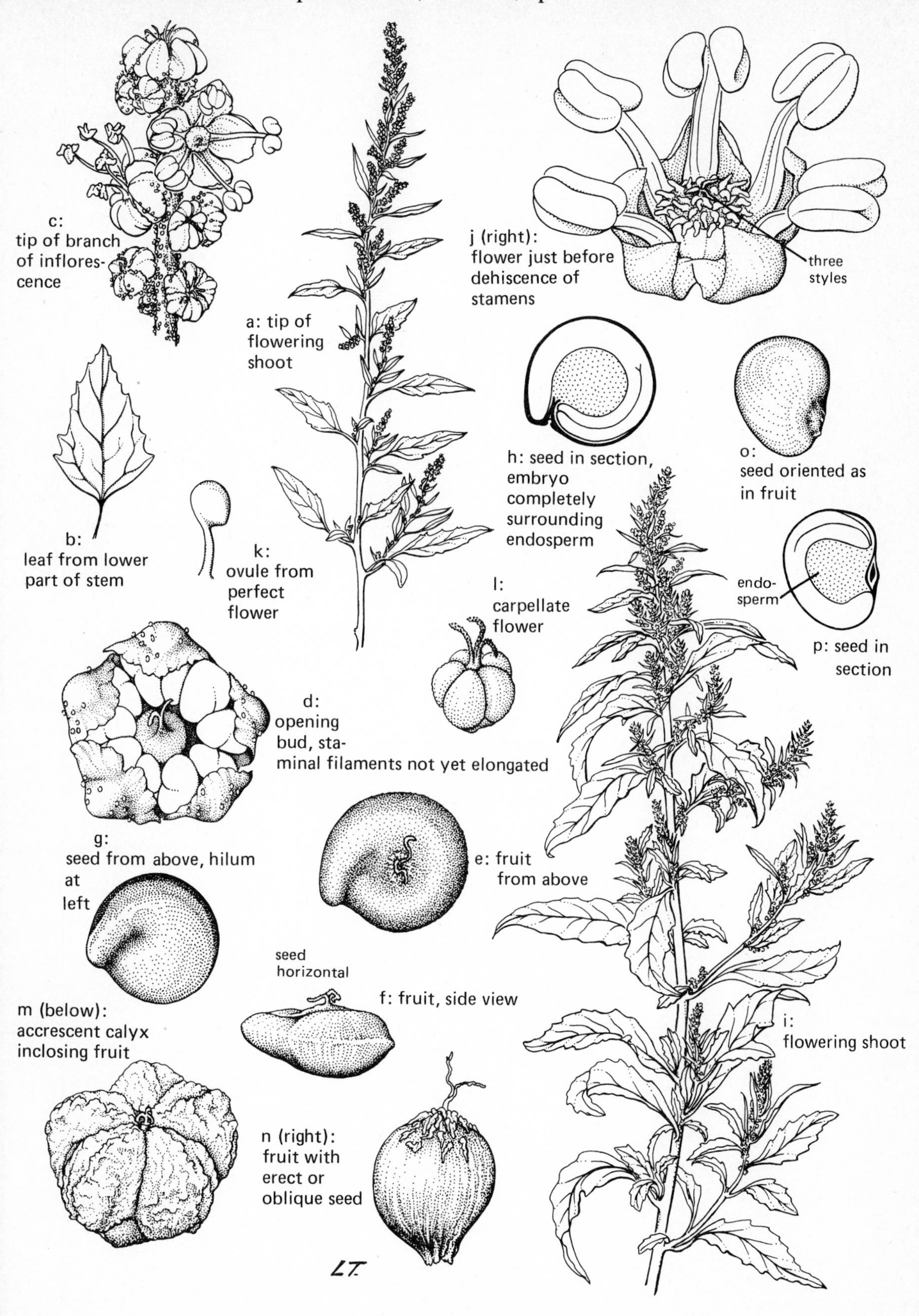

CHENOPODIACEAE: Chenopodium. a-h, C. album; i-p. C. ambrosioides

c:
tip of branch
of inflores-
cence

a: tip of
flowering
shoot

j (right):
flower just before
dehiscence of
stamens

three
styles

h: seed in section,
embryo
completely
surrounding
endosperm

o:
seed oriented as
in fruit

b:
leaf from lower
part of stem

k:
ovule from
perfect
flower

l:
carpellate
flower

endo-
sperm

p: seed in
section

d:
opening
bud, sta-
minal filaments not yet elongated

g:
seed from above, hilum
at
left

e: fruit
from above

seed
horizontal

f: fruit, side view

m (below):
accrescent calyx
inclosing fruit

n (right):
fruit with
erect or
oblique seed

i:
flowering shoot

LT.

AMARANTHACEAE: Amaranthus. a-h, *A. spinosus*; i-j, *A. retroflexus*

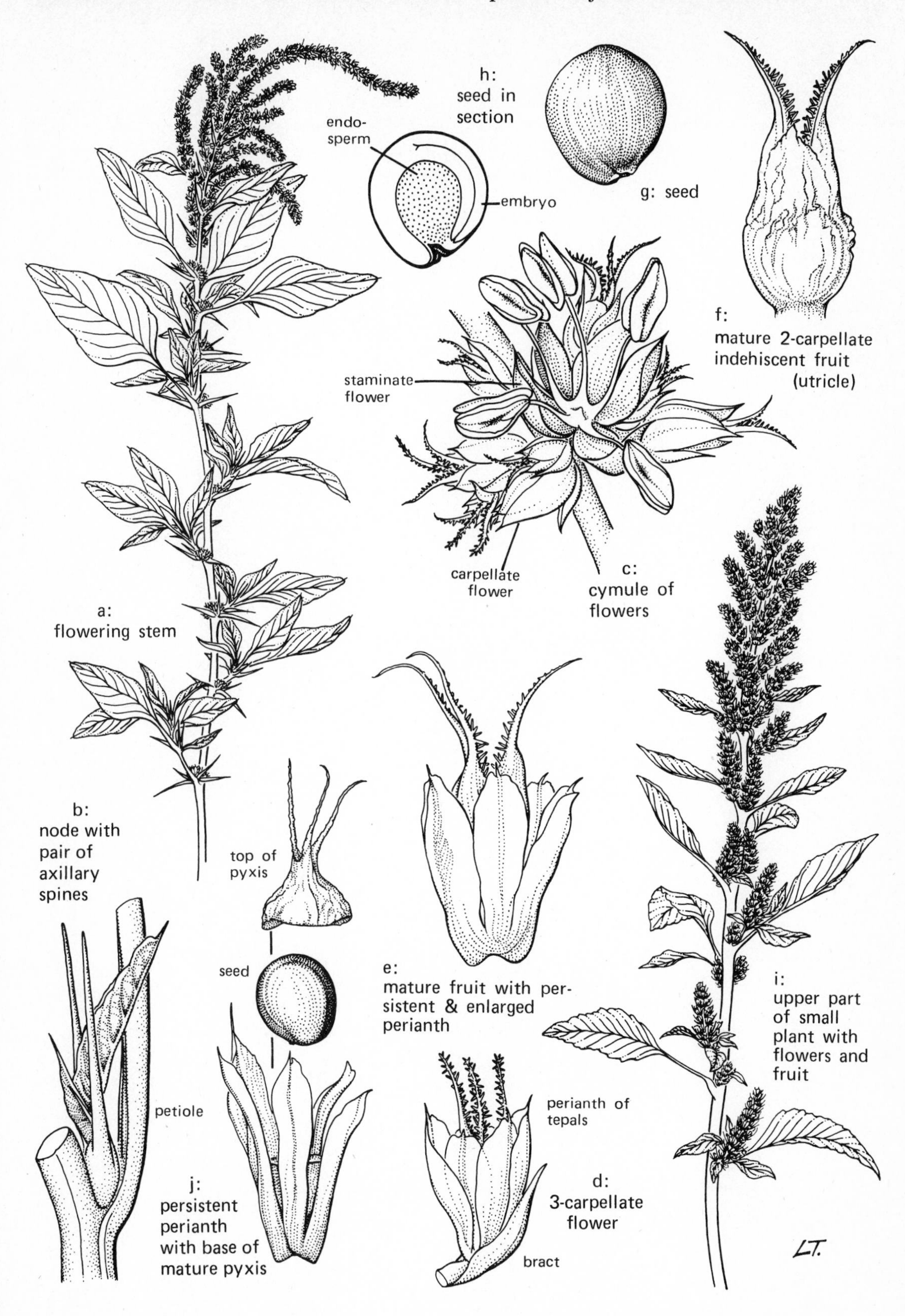

h:
seed in
section

endo-
sperm

embryo

g: seed

f:
mature 2-carpellate
indehiscent fruit
(utricle)

staminate
flower

carpellate
flower

c:
cymule of
flowers

a:
flowering stem

b:
node with
pair of
axillary
spines

top of
pyxis

seed

e:
mature fruit with per-
sistent & enlarged
perianth

i:
upper part
of small
plant with
flowers and
fruit

petiole

perianth of
tepals

j:
persistent
perianth
with base of
mature pyxis

d:
3-carpellate
flower

bract

LT.

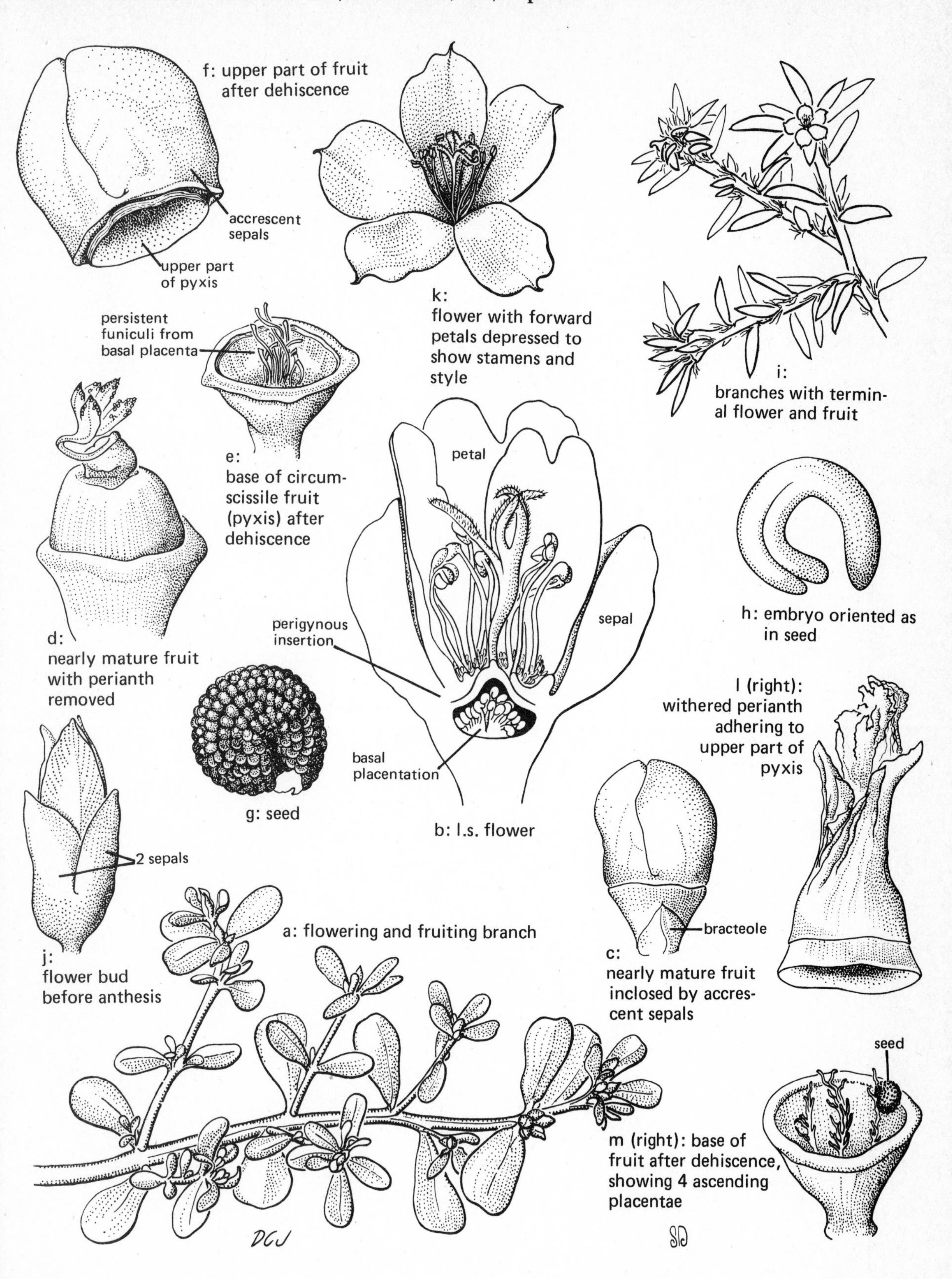

f: upper part of fruit after dehiscence

accrescent sepals

upper part of pyxis

persistent funiculi from basal placenta

k: flower with forward petals depressed to show stamens and style

i: branches with terminal flower and fruit

e: base of circumscissile fruit (pyxis) after dehiscence

petal

h: embryo oriented as in seed

d: nearly mature fruit with perianth removed

perigynous insertion

sepal

l (right): withered perianth adhering to upper part of pyxis

basal placentation

g: seed

b: l.s. flower

2 sepals

j: flower bud before anthesis

a: flowering and fruiting branch

bracteole

c: nearly mature fruit inclosed by accrescent sepals

seed

m (right): base of fruit after dehiscence, showing 4 ascending placentae

DGJ

CARYOPHYLLACEAE: Arenaria. a-i, A. patula; j, A. glabra

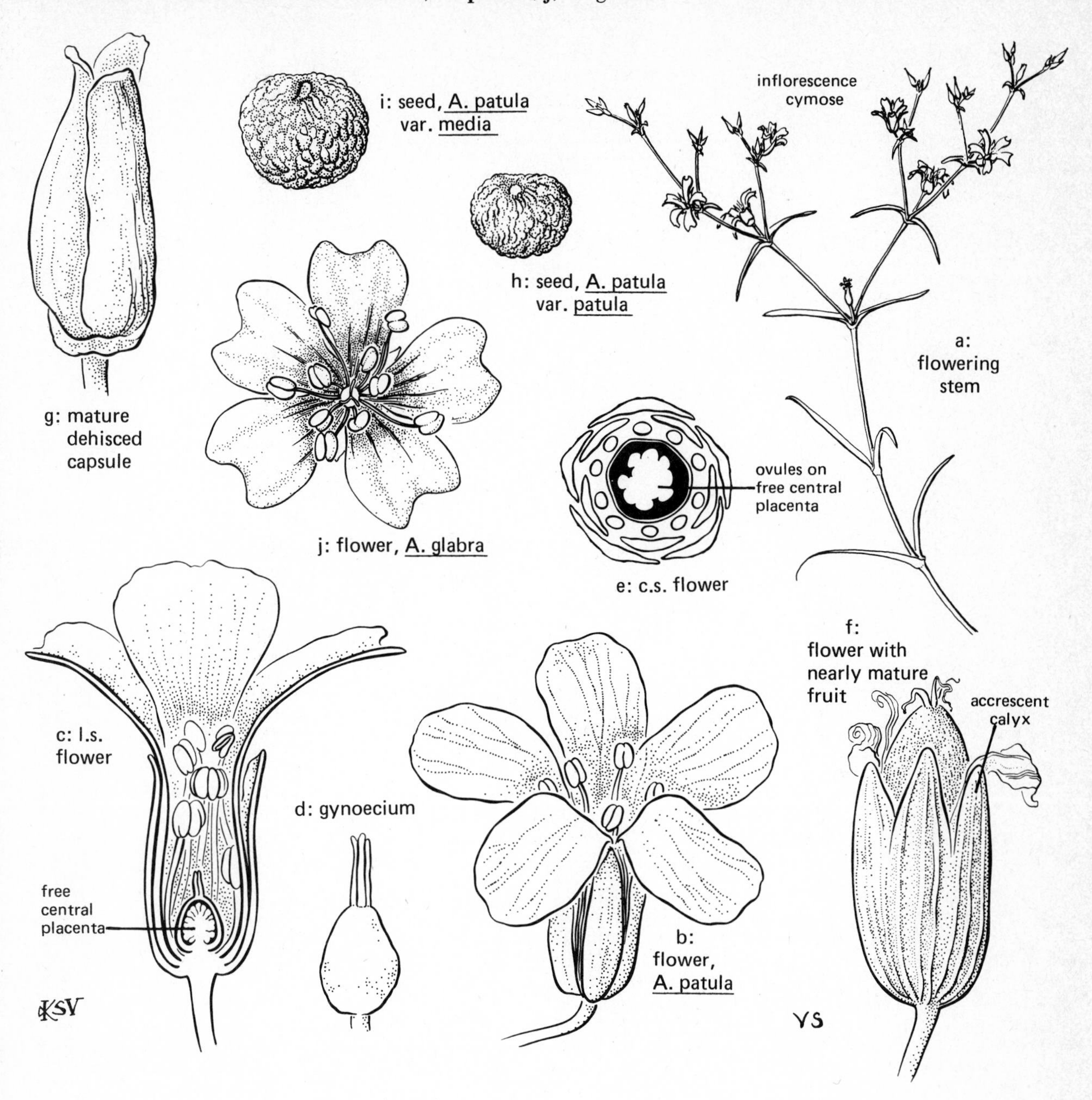

i: seed, <u>A. patula</u> var. <u>media</u>

h: seed, <u>A. patula</u> var. <u>patula</u>

g: mature dehisced capsule

j: flower, <u>A. glabra</u>

inflorescence cymose

a: flowering stem

ovules on free central placenta

e: c.s. flower

f: flower with nearly mature fruit

accrescent calyx

c: l.s. flower

free central placenta

d: gynoecium

b: flower, <u>A. patula</u>

KSV

VS

CARYOPHYLLACEAE: Silene: a-j, S. virginica; k, S. caroliniana; l, S. ovata; m, n, S. antirrhina

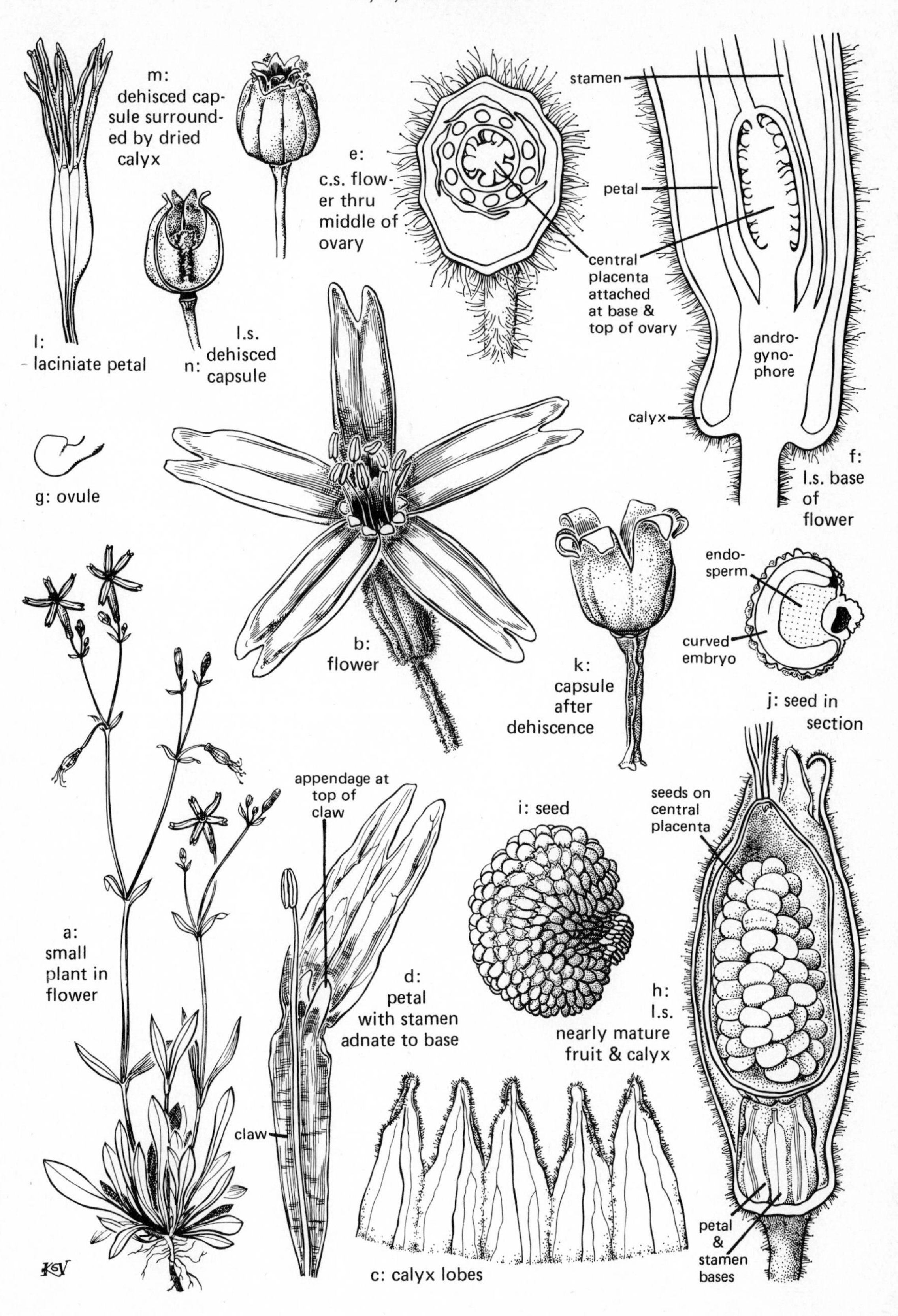

m: dehisced capsule surrounded by dried calyx

l: laciniate petal

n: l.s. dehisced capsule

e: c.s. flower thru middle of ovary

stamen

petal

central placenta attached at base & top of ovary

andro-gyno-phore

calyx

f: l.s. base of flower

g: ovule

b: flower

k: capsule after dehiscence

endo-sperm

curved embryo

j: seed in section

a: small plant in flower

appendage at top of claw

claw

d: petal with stamen adnate to base

i: seed

c: calyx lobes

seeds on central placenta

h: l.s. nearly mature fruit & calyx

petal & stamen bases

NYMPHAEACEAE: Nelumbo. a-m, N. lutea

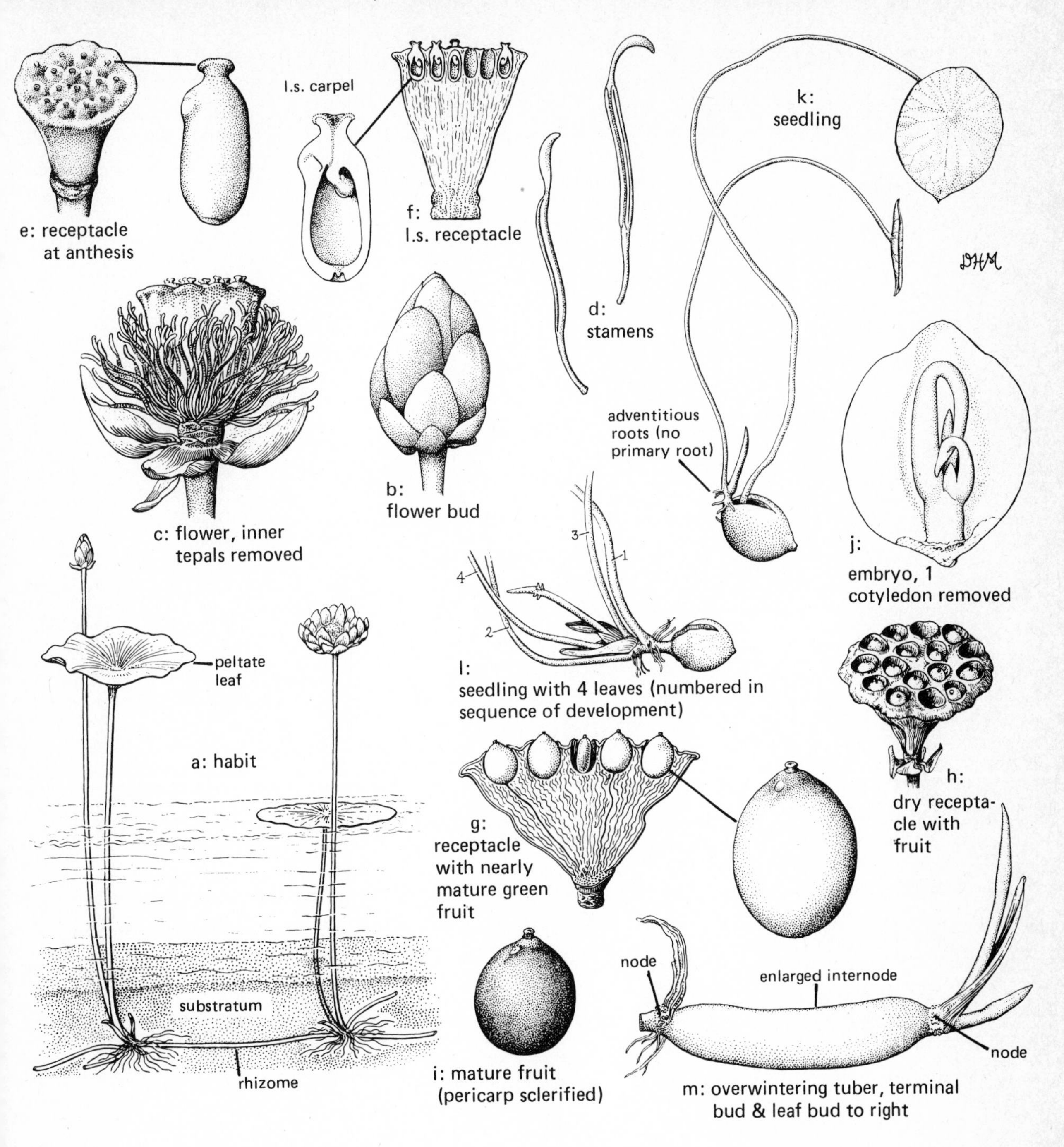

e: receptacle at anthesis

l.s. carpel

f: l.s. receptacle

d: stamens

k: seedling

DHM

adventitious roots (no primary root)

j: embryo, 1 cotyledon removed

c: flower, inner tepals removed

b: flower bud

peltate leaf

a: habit

l: seedling with 4 leaves (numbered in sequence of development)

g: receptacle with nearly mature green fruit

h: dry receptacle with fruit

substratum

rhizome

i: mature fruit (pericarp sclerified)

node

enlarged internode

node

m: overwintering tuber, terminal bud & leaf bud to right

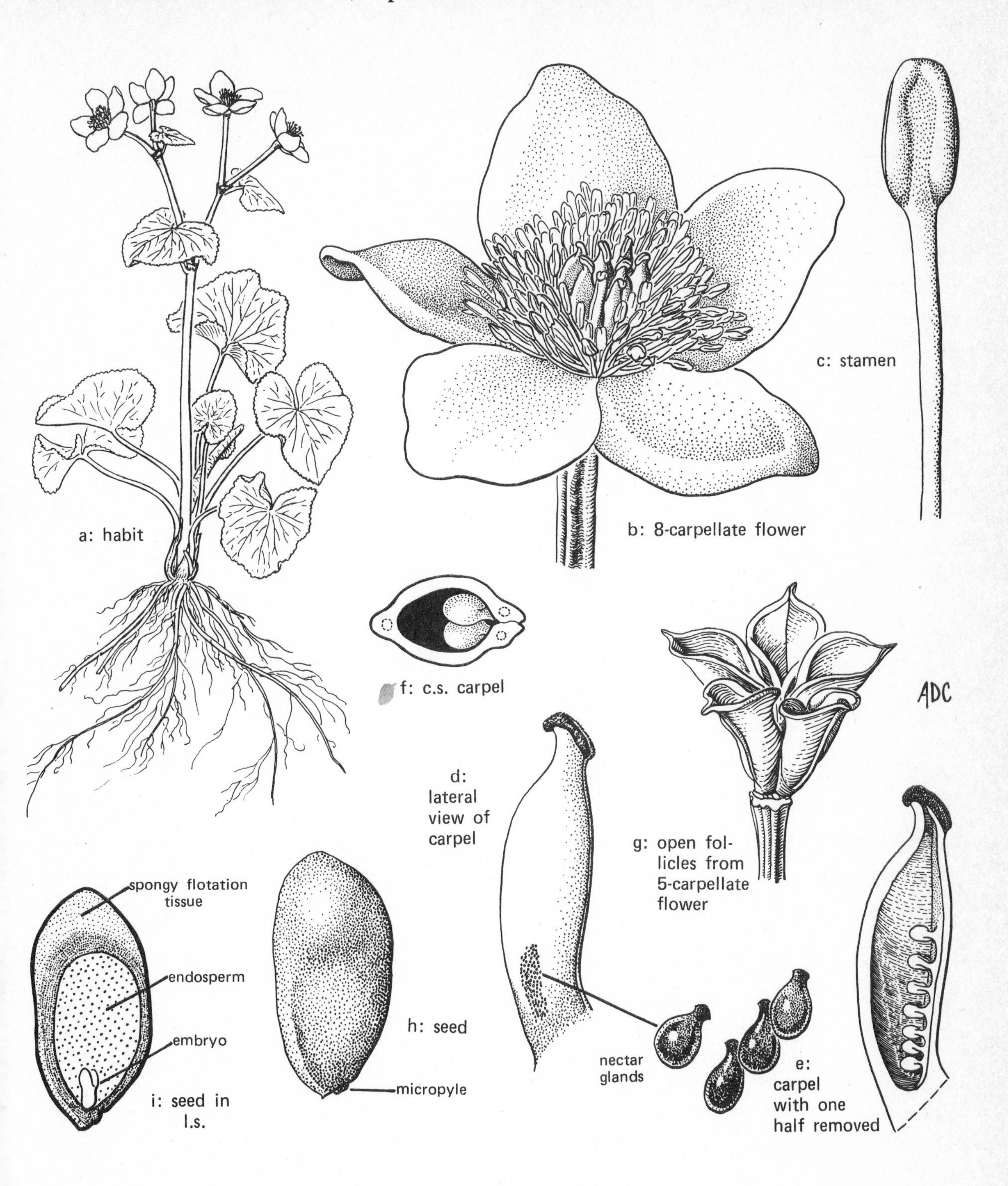

a: habit

b: 8-carpellate flower

c: stamen

f: c.s. carpel

d: lateral view of carpel

g: open follicles from 5-carpellate flower

ADC

spongy flotation tissue

endosperm

embryo

i: seed in l.s.

h: seed

micropyle

nectar glands

e: carpel with one half removed

RANUNCULACEAE: Thalictrum. a-g, T. thalictroides; h-l, T. clavatum; m-p, T. dioicum

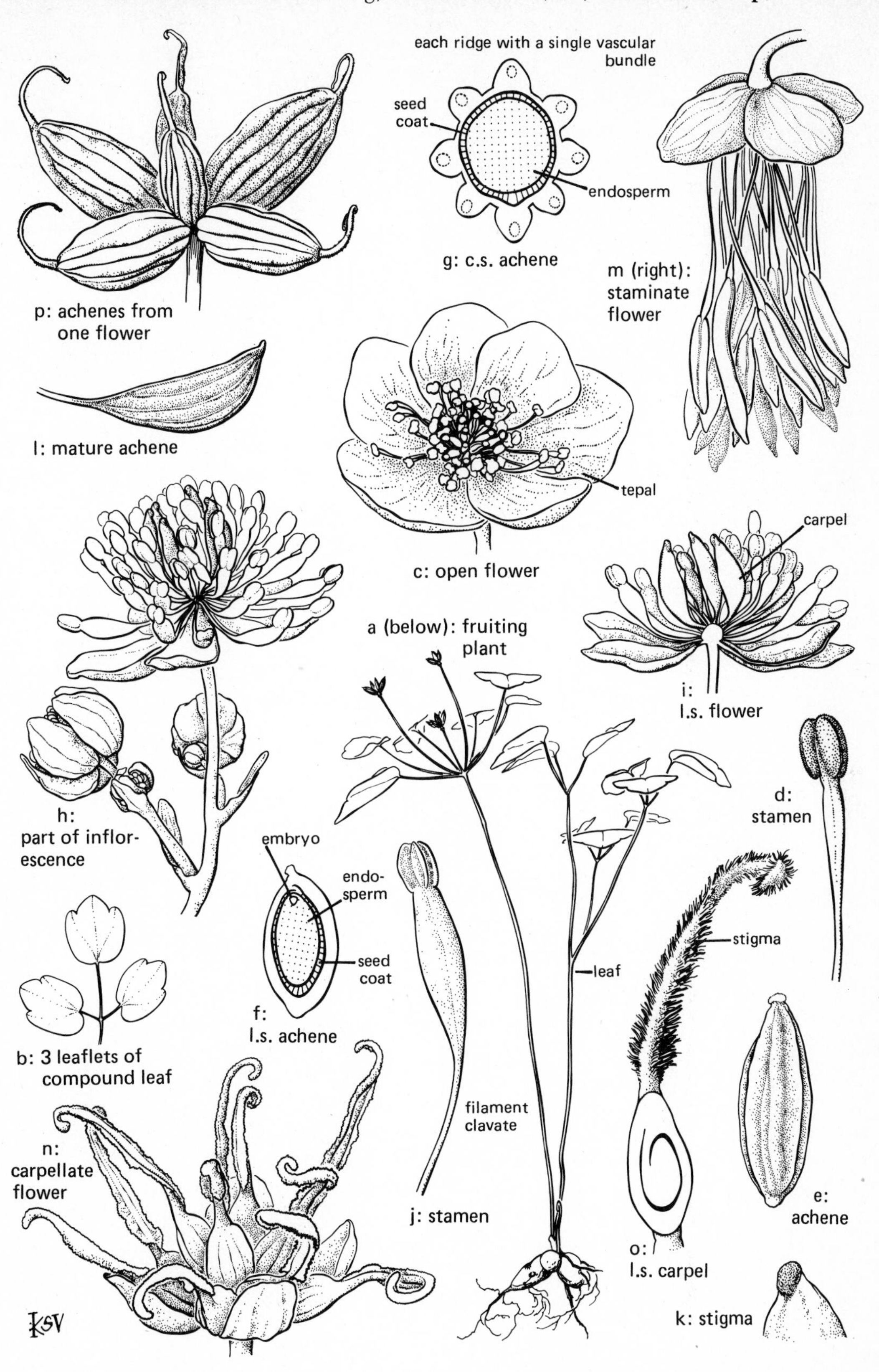

each ridge with a single vascular bundle

seed coat

endosperm

g: c.s. achene

m (right): staminate flower

p: achenes from one flower

l: mature achene

tepal

c: open flower

carpel

a (below): fruiting plant

i: l.s. flower

h: part of inflorescence

embryo

endosperm

seed coat

f: l.s. achene

d: stamen

stigma

leaf

b: 3 leaflets of compound leaf

n: carpellate flower

filament clavate

j: stamen

o: l.s. carpel

e: achene

k: stigma

KSV

30

RANUNCULACEAE: Aquilegia. a-m, A. canadensis

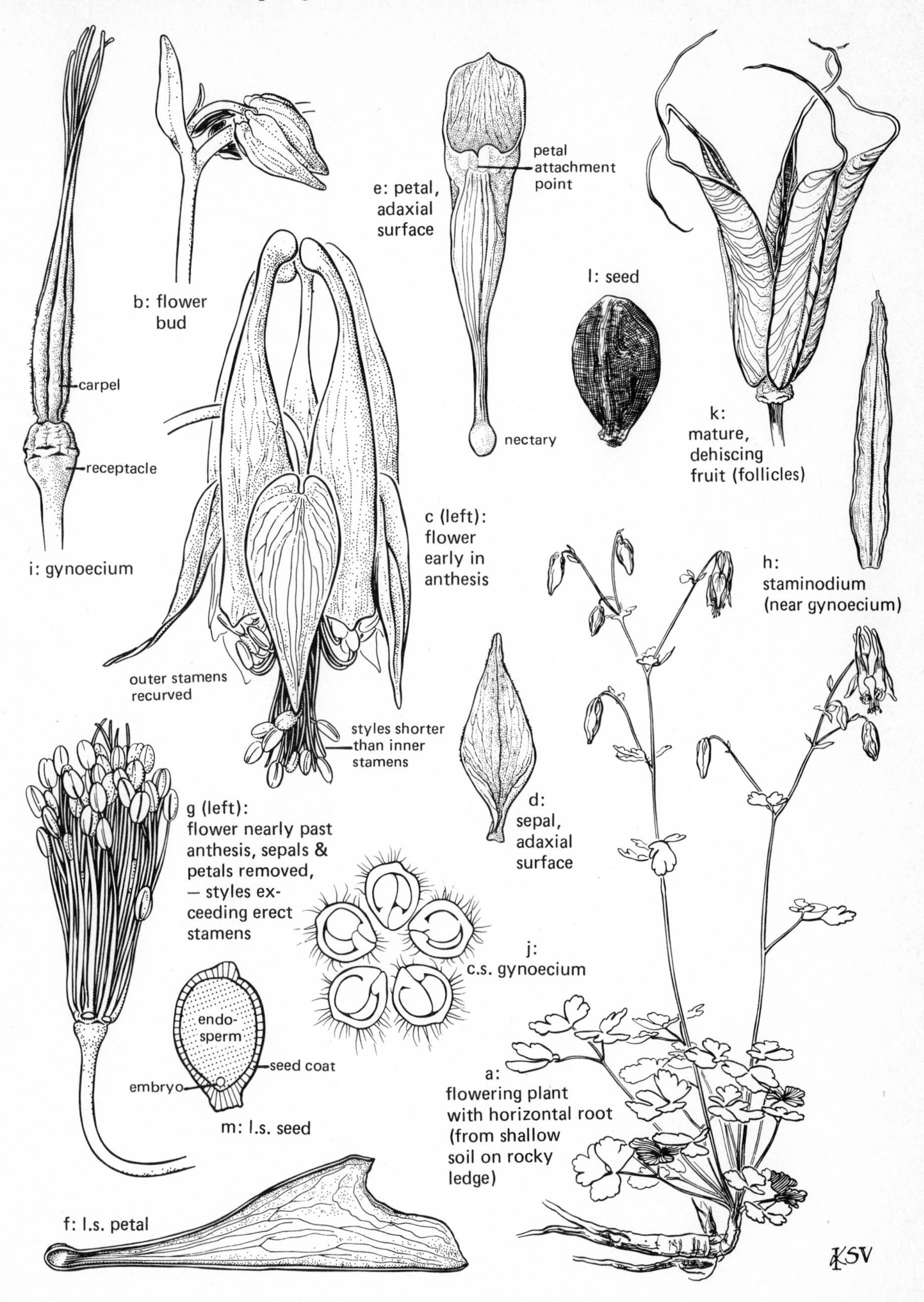

b: flower bud

e: petal, adaxial surface

petal attachment point

l: seed

carpel

receptacle

i: gynoecium

k: mature, dehiscing fruit (follicles)

nectary

c (left): flower early in anthesis

h: staminodium (near gynoecium)

outer stamens recurved

styles shorter than inner stamens

d: sepal, adaxial surface

g (left): flower nearly past anthesis, sepals & petals removed, — styles exceeding erect stamens

j: c.s. gynoecium

endo-sperm

embryo

seed coat

m: l.s. seed

a: flowering plant with horizontal root (from shallow soil on rocky ledge)

f: l.s. petal

KSV

31

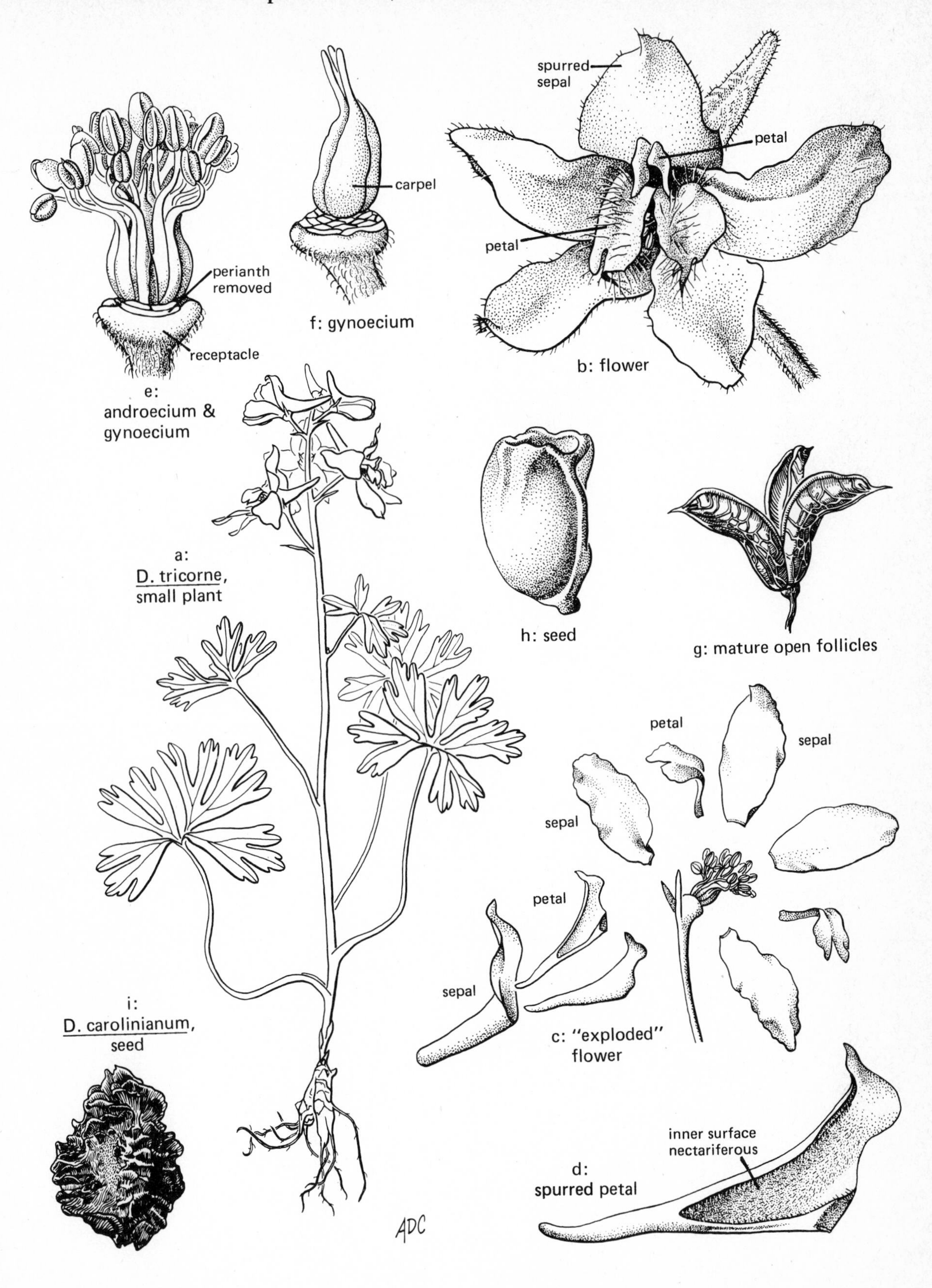

e: androecium & gynoecium

perianth removed

receptacle

f: gynoecium

carpel

b: flower

spurred sepal

petal

petal

a: D. tricorne, small plant

h: seed

g: mature open follicles

petal

sepal

sepal

petal

sepal

petal

sepal

c: "exploded" flower

i: D. carolinianum, seed

d: spurred petal

inner surface nectariferous

ADC

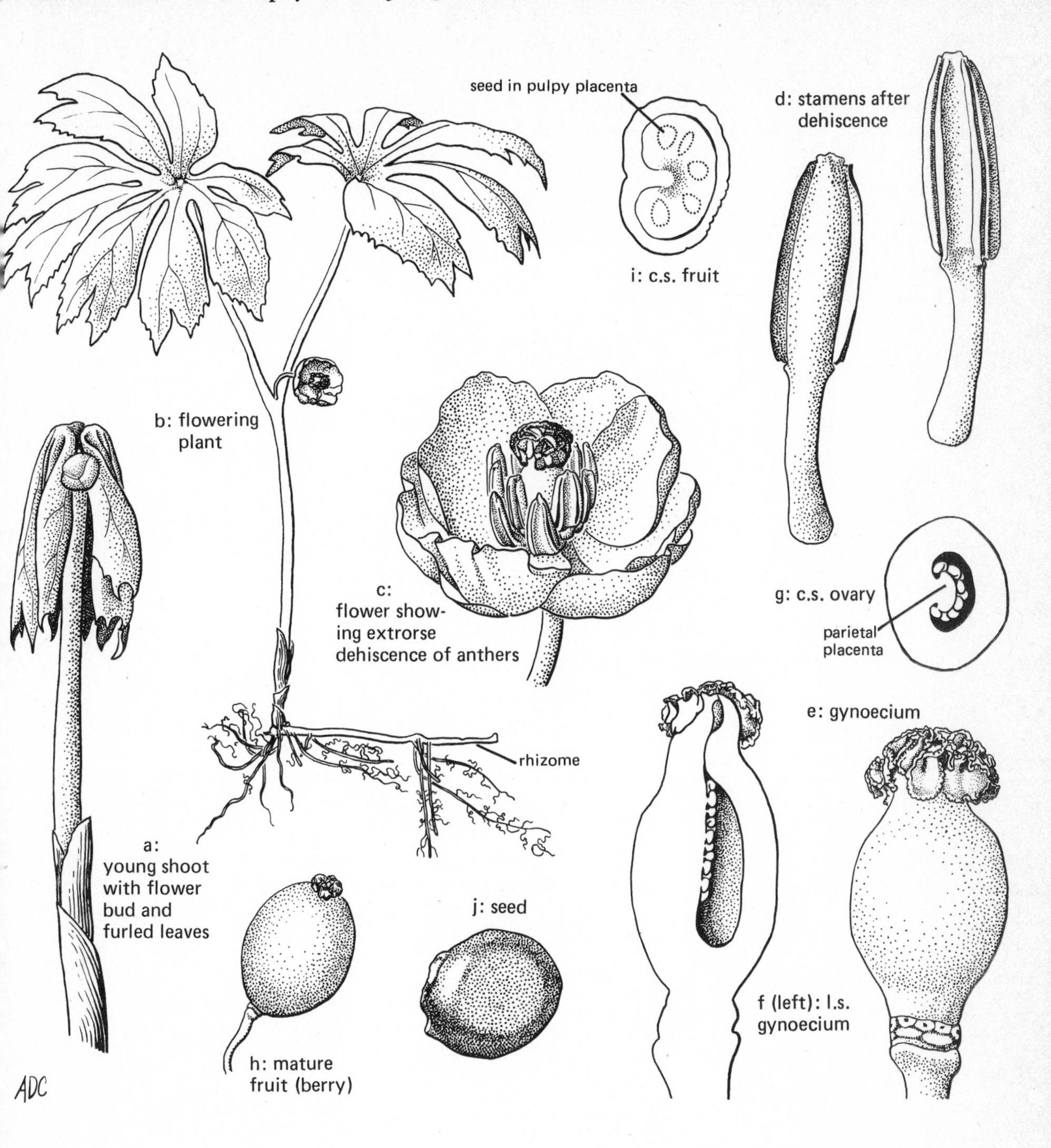

seed in pulpy placenta

i: c.s. fruit

d: stamens after dehiscence

b: flowering plant

c: flower showing extrorse dehiscence of anthers

g: c.s. ovary

parietal placenta

e: gynoecium

rhizome

a: young shoot with flower bud and furled leaves

j: seed

h: mature fruit (berry)

f (left): l.s. gynoecium

ADC

33

MAGNOLIACEAE: Magnolia. a-i, M. virginiana; j-l, M. grandiflora; m, M. tripetala; n-q, M. acuminata

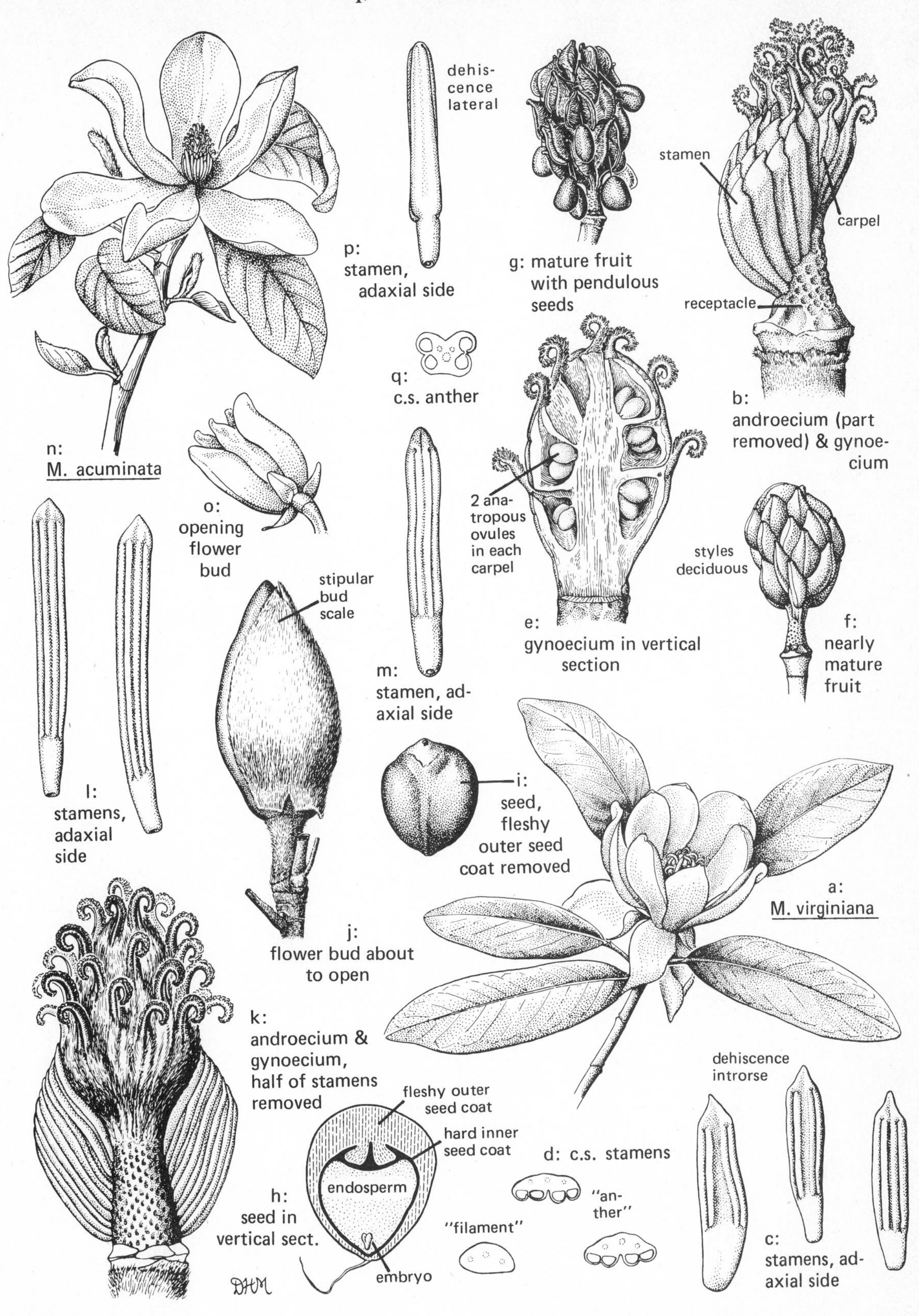

dehis-
cence
lateral

p:
stamen,
adaxial side

g: mature fruit
with pendulous
seeds

stamen

carpel

receptacle

b:
androecium (part
removed) & gynoe-
cium

q:
c.s. anther

2 ana-
tropous
ovules
in each
carpel

styles
deciduous

e:
gynoecium in vertical
section

f:
nearly
mature
fruit

n:
M. acuminata

o:
opening
flower
bud

stipular
bud
scale

m:
stamen, ad-
axial side

i:
seed,
fleshy
outer seed
coat removed

a:
M. virginiana

l:
stamens,
adaxial
side

j:
flower bud about
to open

k:
androecium &
gynoecium,
half of stamens
removed

fleshy outer
seed coat

hard inner
seed coat

dehiscence
introrse

endosperm

d: c.s. stamens

"an-
ther"

h:
seed in
vertical sect.

"filament"

embryo

c:
stamens, ad-
axial side

DHM

MAGNOLIACEAE: Liriodendron. a-l, L. tulipifera

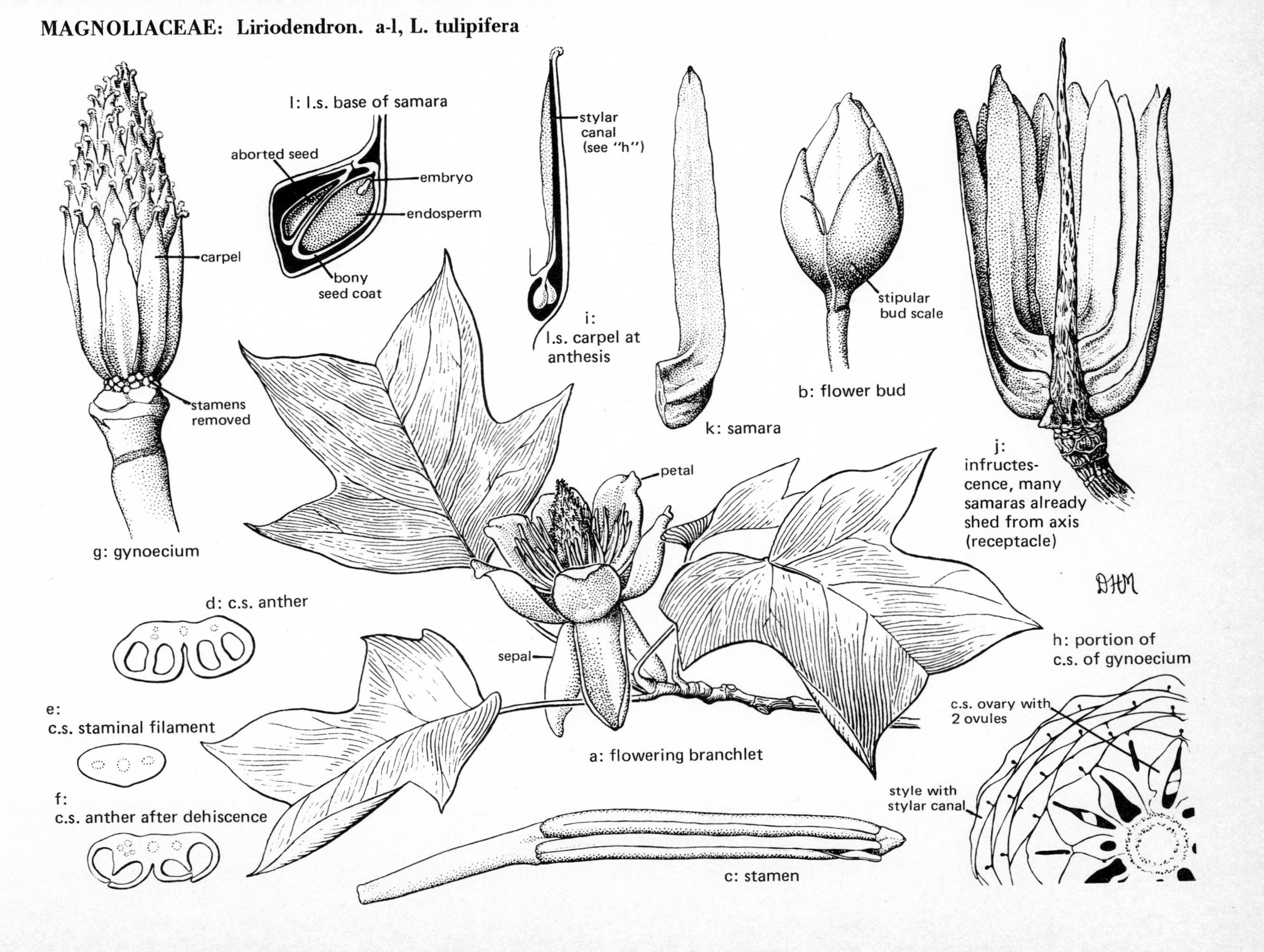

j: infructescence, many samaras already shed from axis (receptacle)

h: portion of c.s. of gynoecium

c.s. ovary with 2 ovules

style with stylar canal

stipular bud scale

b: flower bud

k: samara

c: stamen

petal

a: flowering branchlet

sepal

stylar canal (see "h")

i: l.s. carpel at anthesis

embryo

endosperm

bony seed coat

l: l.s. base of samara

aborted seed

carpel

stamens removed

g: gynoecium

d: c.s. anther

e: c.s. staminal filament

f: c.s. anther after dehiscence

ANNONACEAE: Asimina. a-i, A. triloba; j-l, A. incana; m, A. obovata

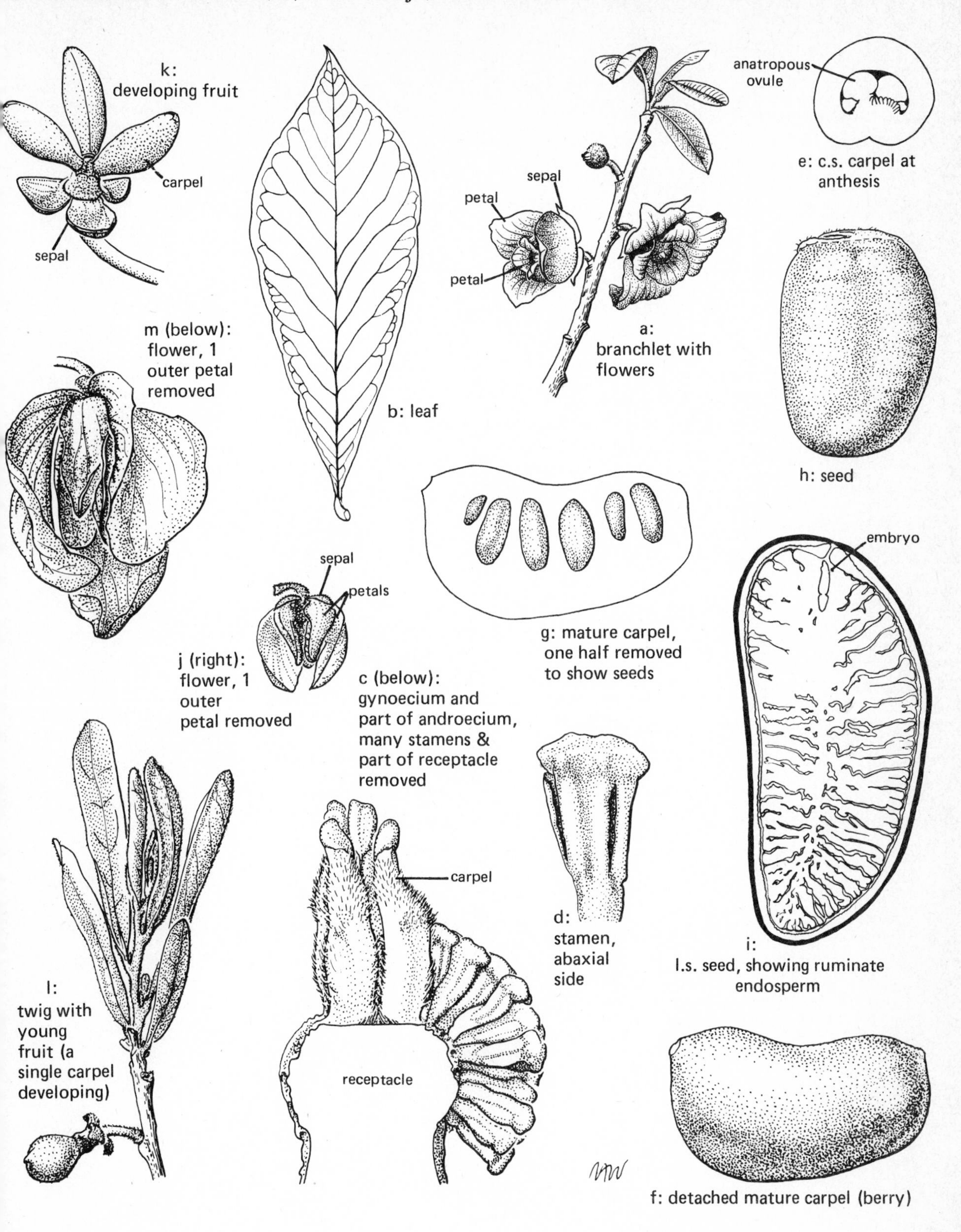

k: developing fruit

carpel

sepal

m (below): flower, 1 outer petal removed

b: leaf

petal

sepal

petal

a: branchlet with flowers

anatropous ovule

e: c.s. carpel at anthesis

h: seed

sepal

petals

j (right): flower, 1 outer petal removed

c (below): gynoecium and part of androecium, many stamens & part of receptacle removed

g: mature carpel, one half removed to show seeds

embryo

carpel

d: stamen, abaxial side

i: l.s. seed, showing ruminate endosperm

l: twig with young fruit (a single carpel developing)

receptacle

f: detached mature carpel (berry)

LAURACEAE: Sassafras. a-h, S. albidum var. molle

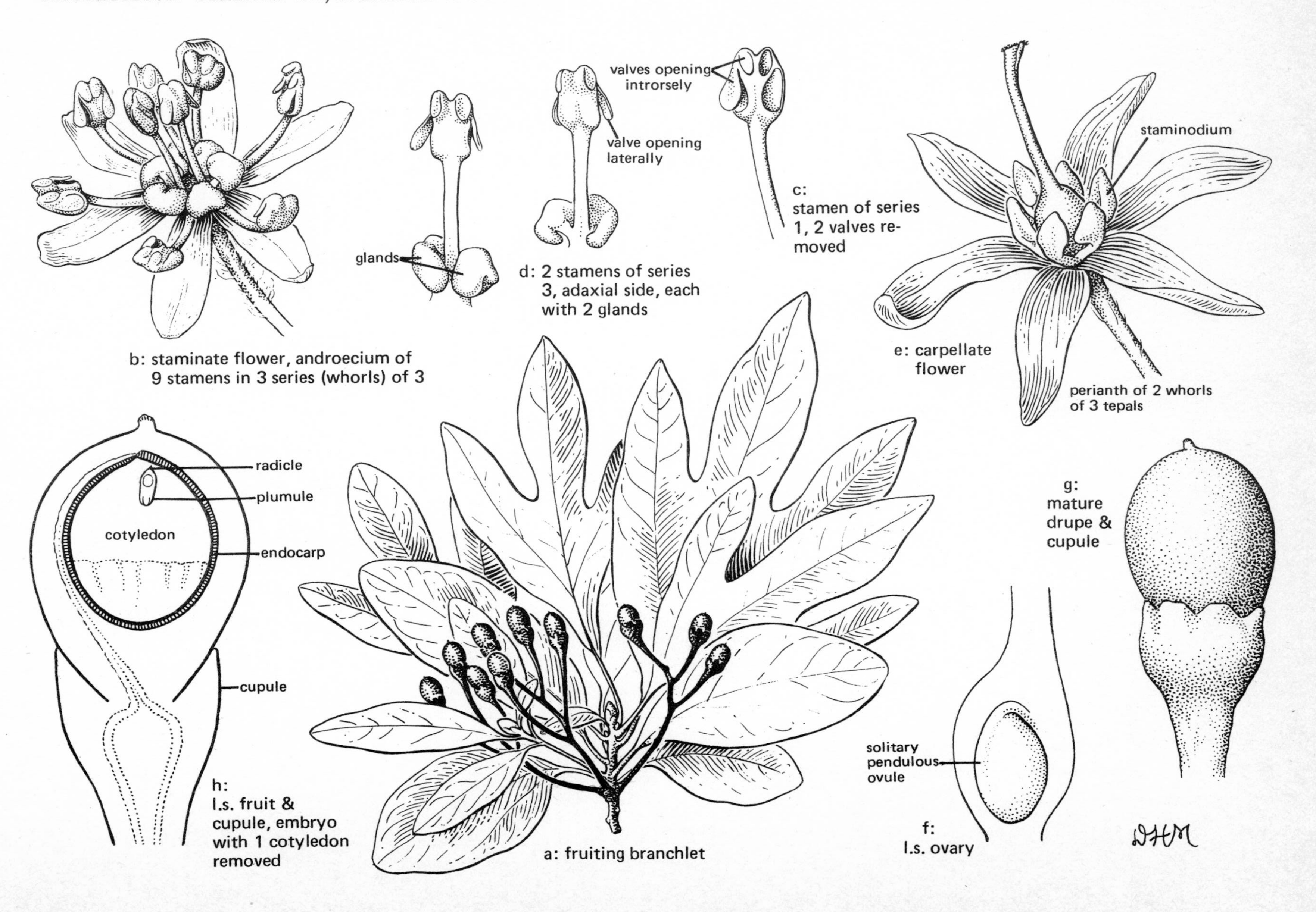

staminodium

perianth of 2 whorls of 3 tepals

e: carpellate flower

g: mature drupe & cupule

solitary pendulous ovule

f: l.s. ovary

valves opening introrsely

valve opening laterally

c: stamen of series 1, 2 valves removed

d: 2 stamens of series 3, adaxial side, each with 2 glands

glands

b: staminate flower, androecium of 9 stamens in 3 series (whorls) of 3

a: fruiting branchlet

radicle

plumule

endocarp

cotyledon

cupule

h: l.s. fruit & cupule, embryo with 1 cotyledon removed

LAURACEAE: Lindera. a-i, L. benzoin

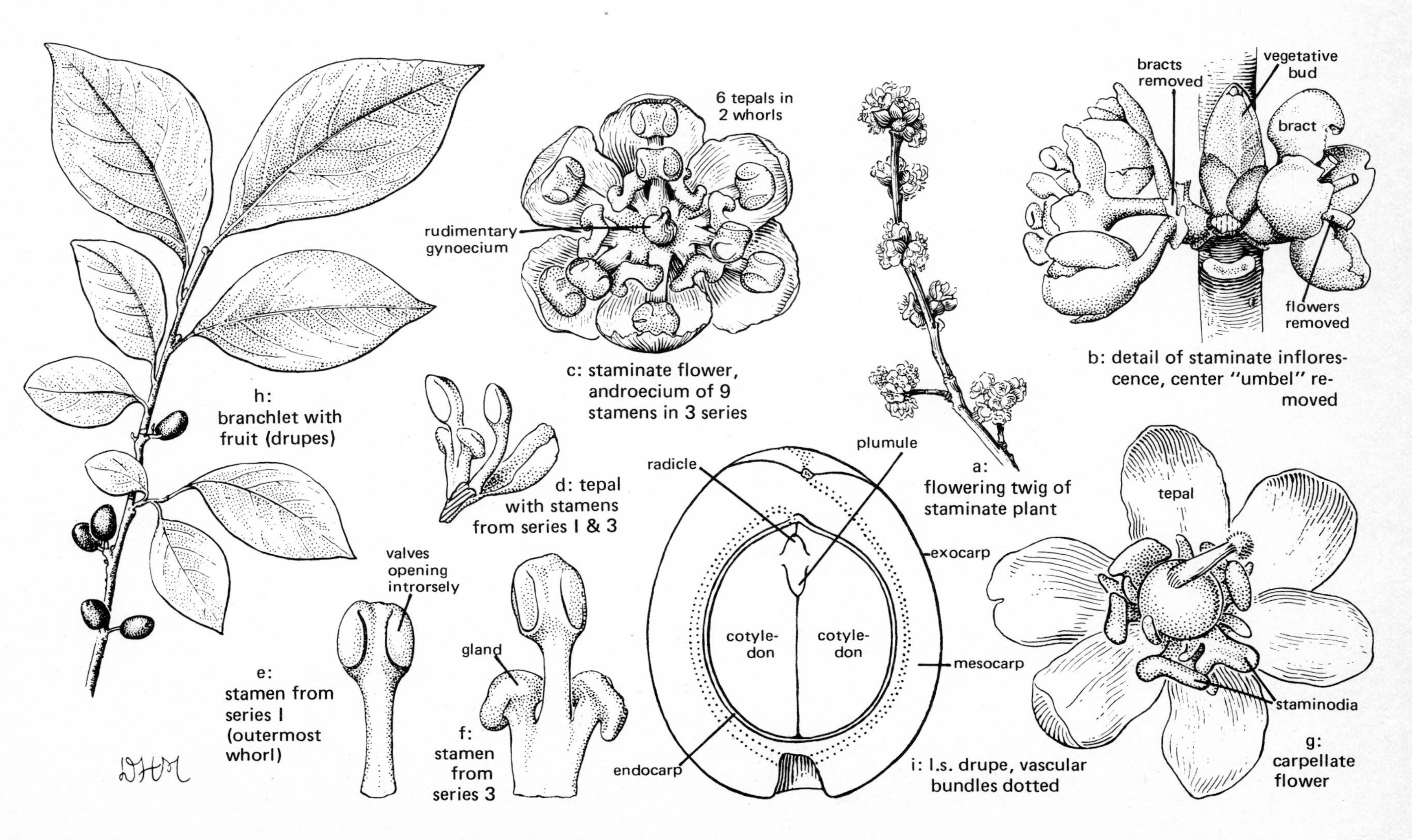

a: flowering twig of staminate plant

b: detail of staminate inflorescence, center "umbel" removed

vegetative bud

bract

flowers removed

bracts removed

tepal

staminodia

g: carpellate flower

6 tepals in 2 whorls

c: staminate flower, androecium of 9 stamens in 3 series

rudimentary gynoecium

d: tepal with stamens from series 1 & 3

gland

f: stamen from series 3

valves opening introrsely

e: stamen from series 1 (outermost whorl)

exocarp

mesocarp

plumule

cotyle-don

cotyle-don

radicle

endocarp

i: l.s. drupe, vascular bundles dotted

h: branchlet with fruit (drupes)

PAPAVERACEAE: Stylophorum. a-h, S. diphyllum

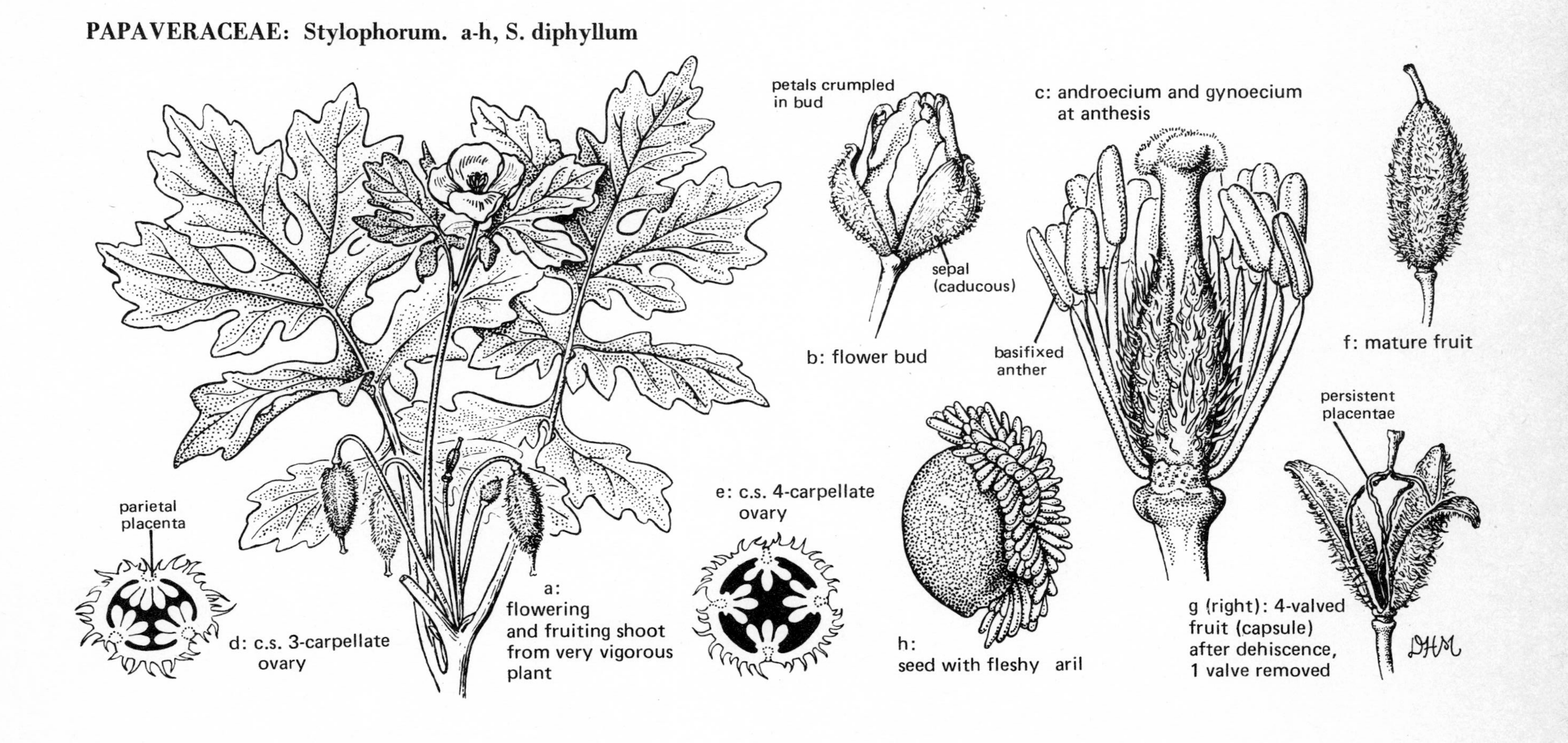

f: mature fruit

persistent placentae

g (right): 4-valved fruit (capsule) after dehiscence, 1 valve removed

c: androecium and gynoecium at anthesis

petals crumpled in bud

sepal (caducous)

basifixed anther

b: flower bud

h: seed with fleshy aril

e: c.s. 4-carpellate ovary

a: flowering and fruiting shoot from very vigorous plant

d: c.s. 3-carpellate ovary

parietal placenta

39

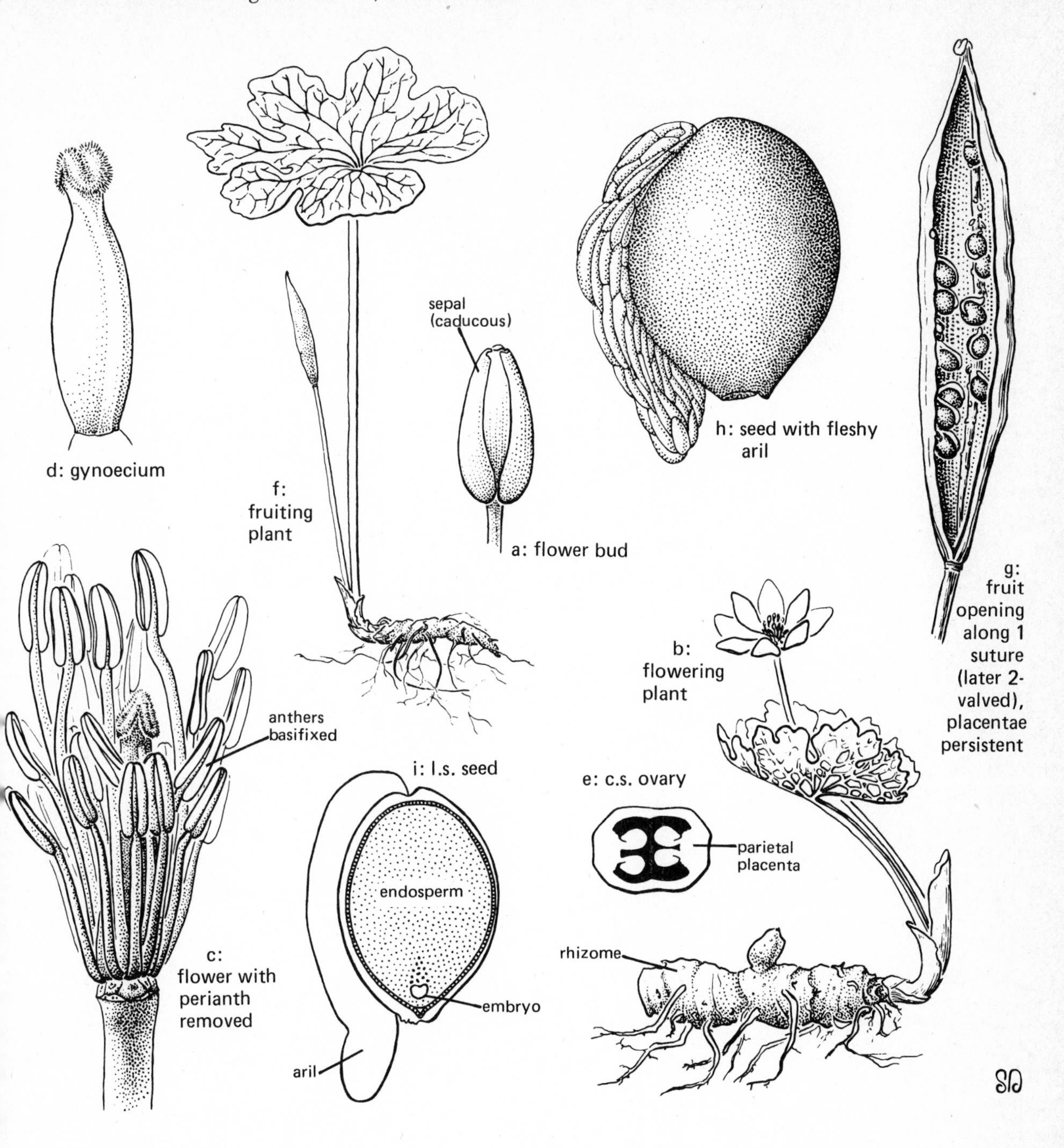

d: gynoecium

f: fruiting plant

sepal (caducous)

a: flower bud

h: seed with fleshy aril

g: fruit opening along 1 suture (later 2-valved), placentae persistent

anthers basifixed

b: flowering plant

i: l.s. seed

e: c.s. ovary

endosperm

embryo

aril

parietal placenta

rhizome

c: flower with perianth removed

40

FUMARIACEAE: Corydalis. a-l, C. sempervirens

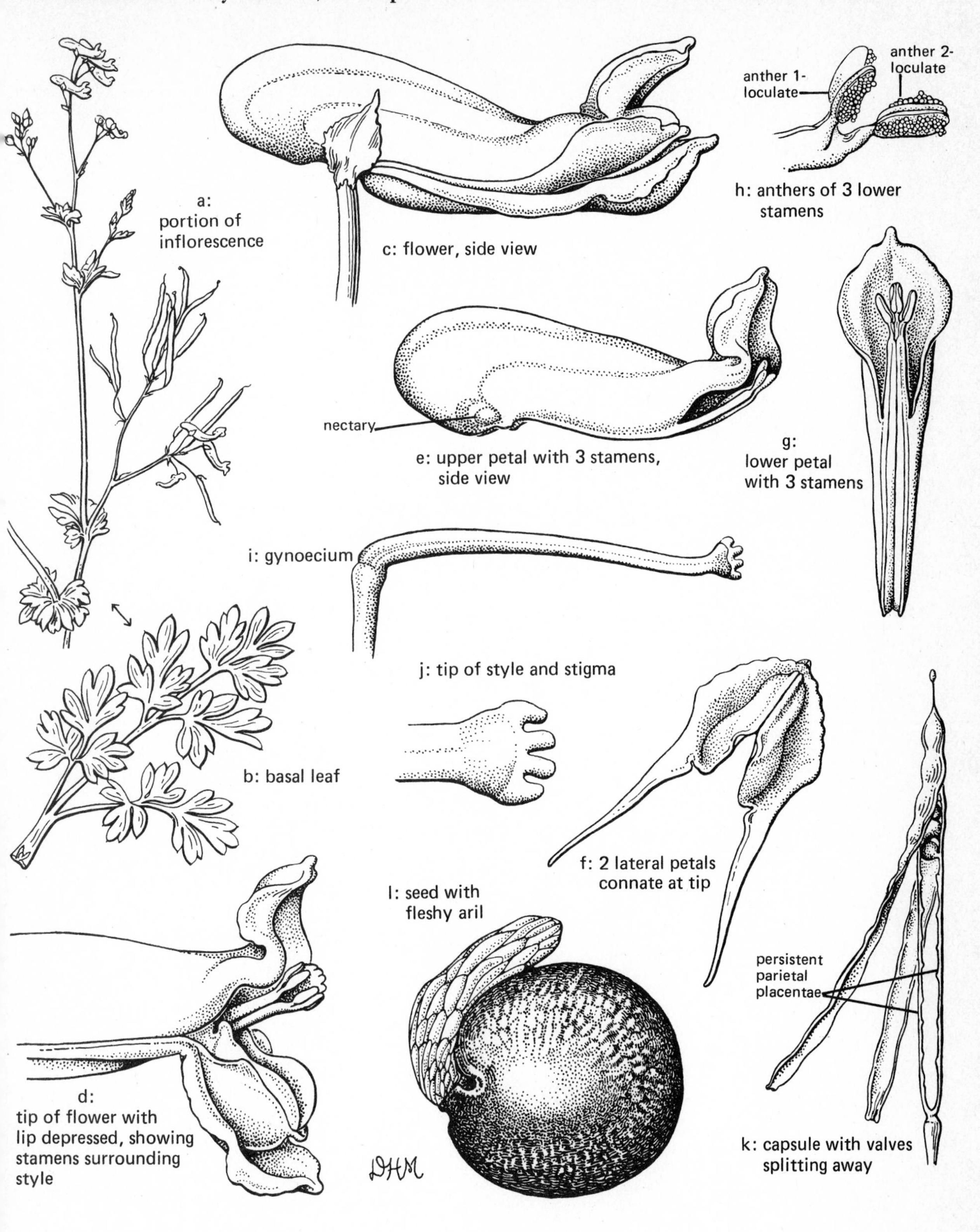

a:
portion of
inflorescence

c: flower, side view

anther 1-
loculate

anther 2-
loculate

h: anthers of 3 lower
stamens

nectary

e: upper petal with 3 stamens,
side view

g:
lower petal
with 3 stamens

i: gynoecium

j: tip of style and stigma

b: basal leaf

l: seed with
fleshy aril

f: 2 lateral petals
connate at tip

persistent
parietal
placentae

d:
tip of flower with
lip depressed, showing
stamens surrounding
style

k: capsule with valves
splitting away

DHM

41

BRASSICACEAE (CRUCIFERAE): a-h, Capsella bursa-pastoris; i-t, fruits, seeds, & embryos of various species

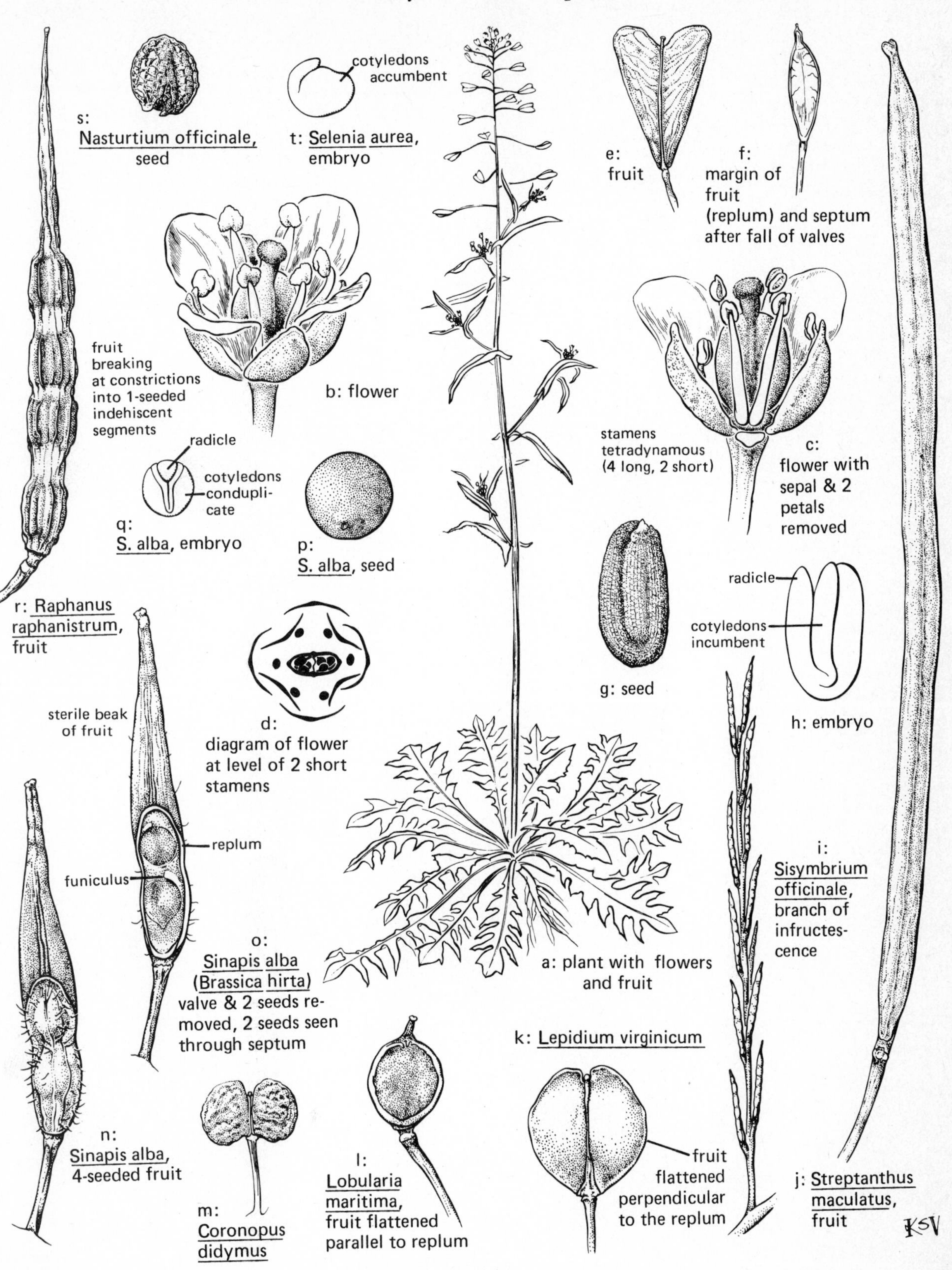

s:
Nasturtium officinale,
seed

t: Selenia aurea,
embryo

cotyledons
accumbent

e:
fruit

f:
margin of
fruit
(replum) and septum
after fall of valves

fruit
breaking
at constrictions
into 1-seeded
indehiscent
segments

b: flower

radicle

cotyledons
conduplicate

q:
S. alba, embryo

p:
S. alba, seed

stamens
tetradynamous
(4 long, 2 short)

c:
flower with
sepal & 2
petals
removed

r: Raphanus
raphanistrum,
fruit

radicle

cotyledons
incumbent

g: seed

h: embryo

sterile beak
of fruit

d:
diagram of flower
at level of 2 short
stamens

replum

funiculus

o:
Sinapis alba
(Brassica hirta)
valve & 2 seeds re-
moved, 2 seeds seen
through septum

i:
Sisymbrium
officinale,
branch of
infructescence

a: plant with flowers
and fruit

k: Lepidium virginicum

n:
Sinapis alba,
4-seeded fruit

m:
Coronopus
didymus

l:
Lobularia
maritima,
fruit flattened
parallel to replum

fruit
flattened
perpendicular
to the replum

j: Streptanthus
maculatus,
fruit

KSV

42

BRASSICACEAE (CRUCIFERAE): a-p, fruits and seeds of various species

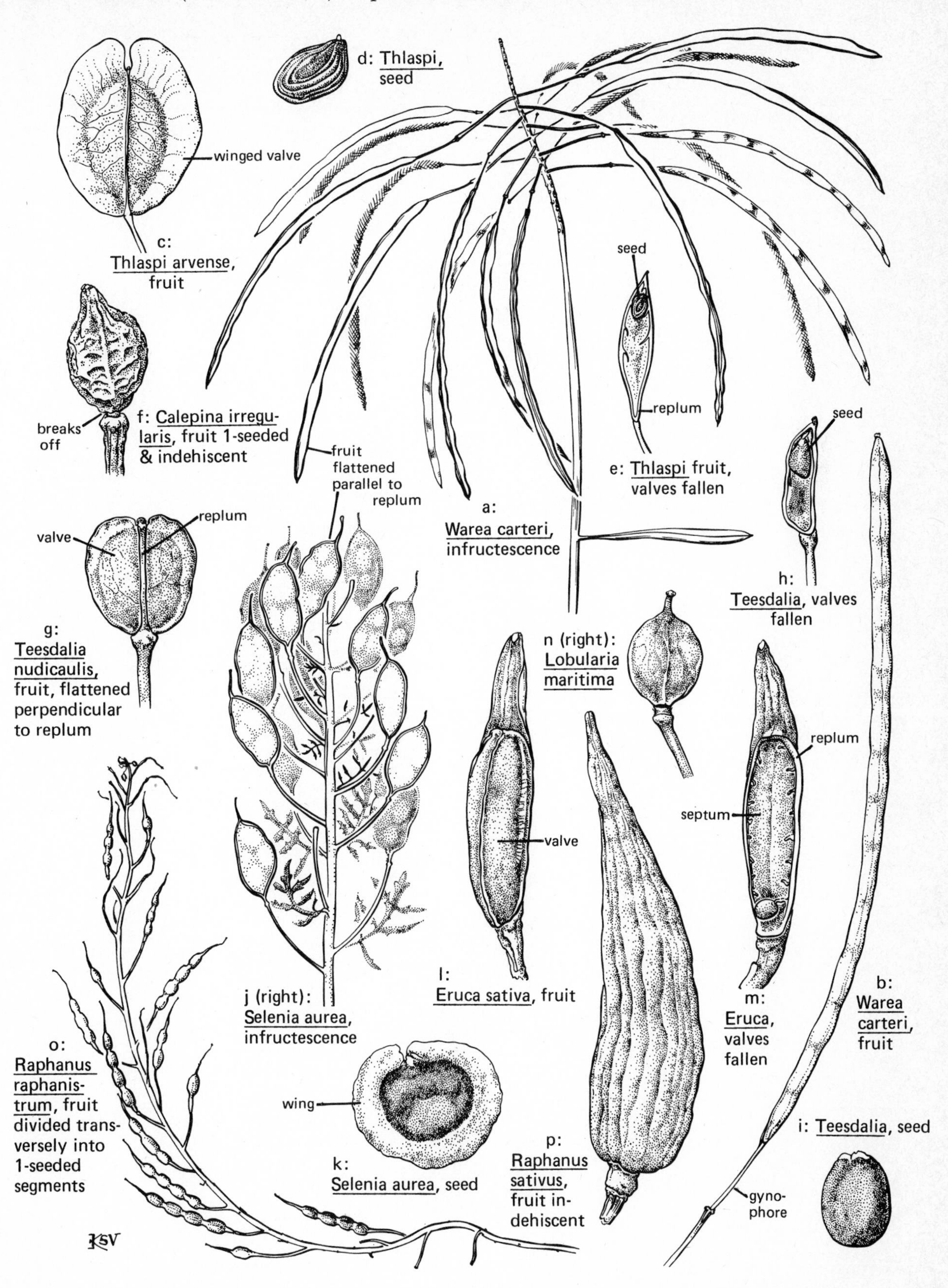

winged valve

d: Thlaspi, seed

c: Thlaspi arvense, fruit

seed

replum

breaks off

f: Calepina irregularis, fruit 1-seeded & indehiscent

fruit flattened parallel to replum

e: Thlaspi fruit, valves fallen

seed

valve

replum

a: Warea carteri, infructescence

h: Teesdalia, valves fallen

g: Teesdalia nudicaulis, fruit, flattened perpendicular to replum

n (right): Lobularia maritima

replum

septum

valve

b: Warea carteri, fruit

j (right): Selenia aurea, infructescence

l: Eruca sativa, fruit

m: Eruca, valves fallen

o: Raphanus raphanistrum, fruit divided transversely into 1-seeded segments

wing

k: Selenia aurea, seed

p: Raphanus sativus, fruit indehiscent

i: Teesdalia, seed

gynophore

KSV

SARRACENIACEAE: Sarracenia. a-h, S. oreophila; i, S. flava; j, S. rubra; k, l, S. purpurea; m, S. psittacina

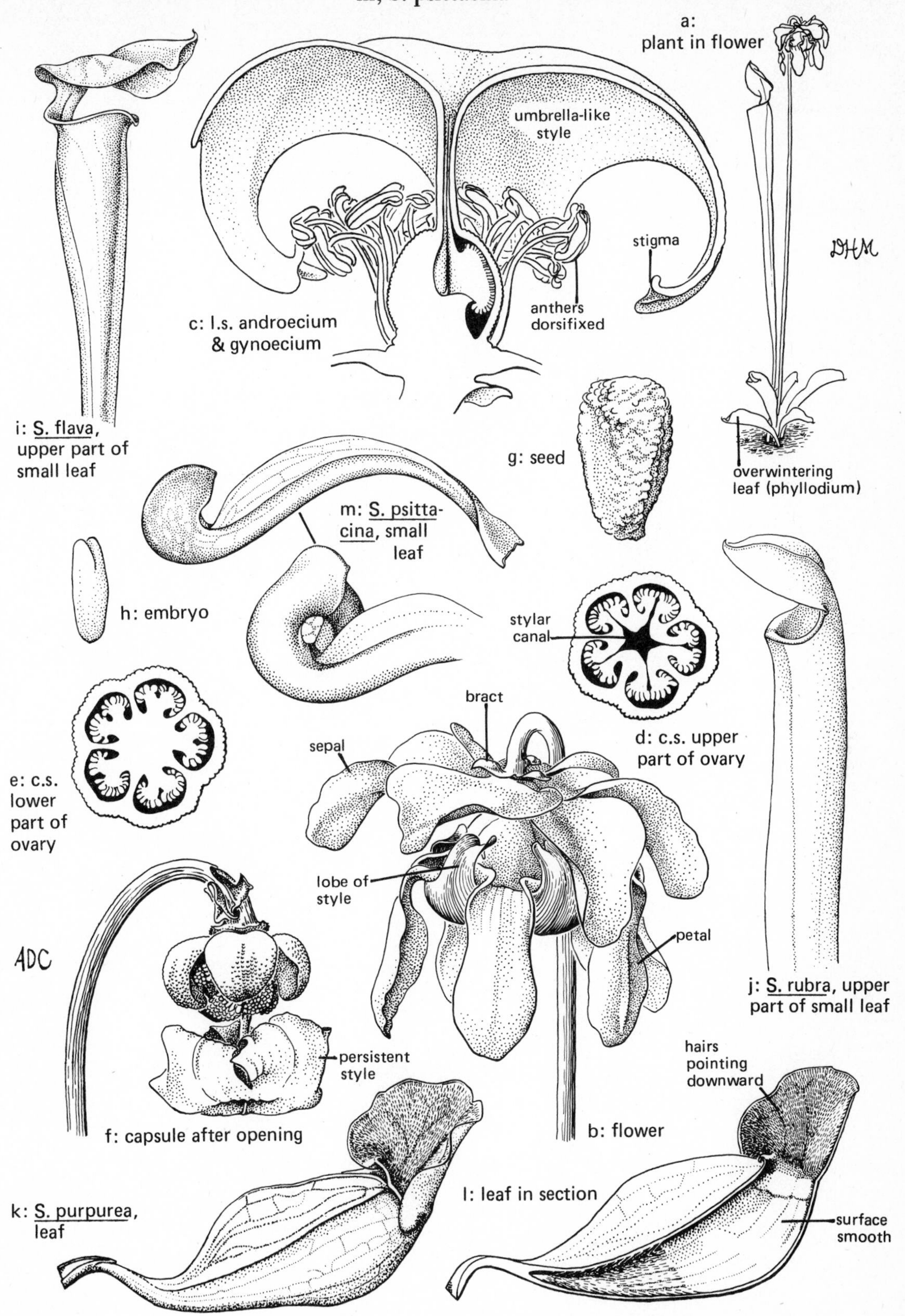

a:
plant in flower

umbrella-like
style

stigma

c: l.s. androecium
& gynoecium

anthers
dorsifixed

DHM

i: S. flava,
upper part of
small leaf

g: seed

overwintering
leaf (phyllodium)

m: S. psitta-
cina, small
leaf

h: embryo

stylar
canal

d: c.s. upper
part of ovary

e: c.s.
lower
part of
ovary

bract

sepal

lobe of
style

petal

j: S. rubra, upper
part of small leaf

ADC

persistent
style

hairs
pointing
downward

f: capsule after opening

b: flower

k: S. purpurea,
leaf

l: leaf in section

surface
smooth

DROSERACEAE: Drosera. a-e, **D. filiformis** var. **tracyi;** f-j, var. **filiformis;** k, l, **D. capillaris;** m, **D. intermedia;** n, **D. rotundifolia;** o, **D. brevifolia**

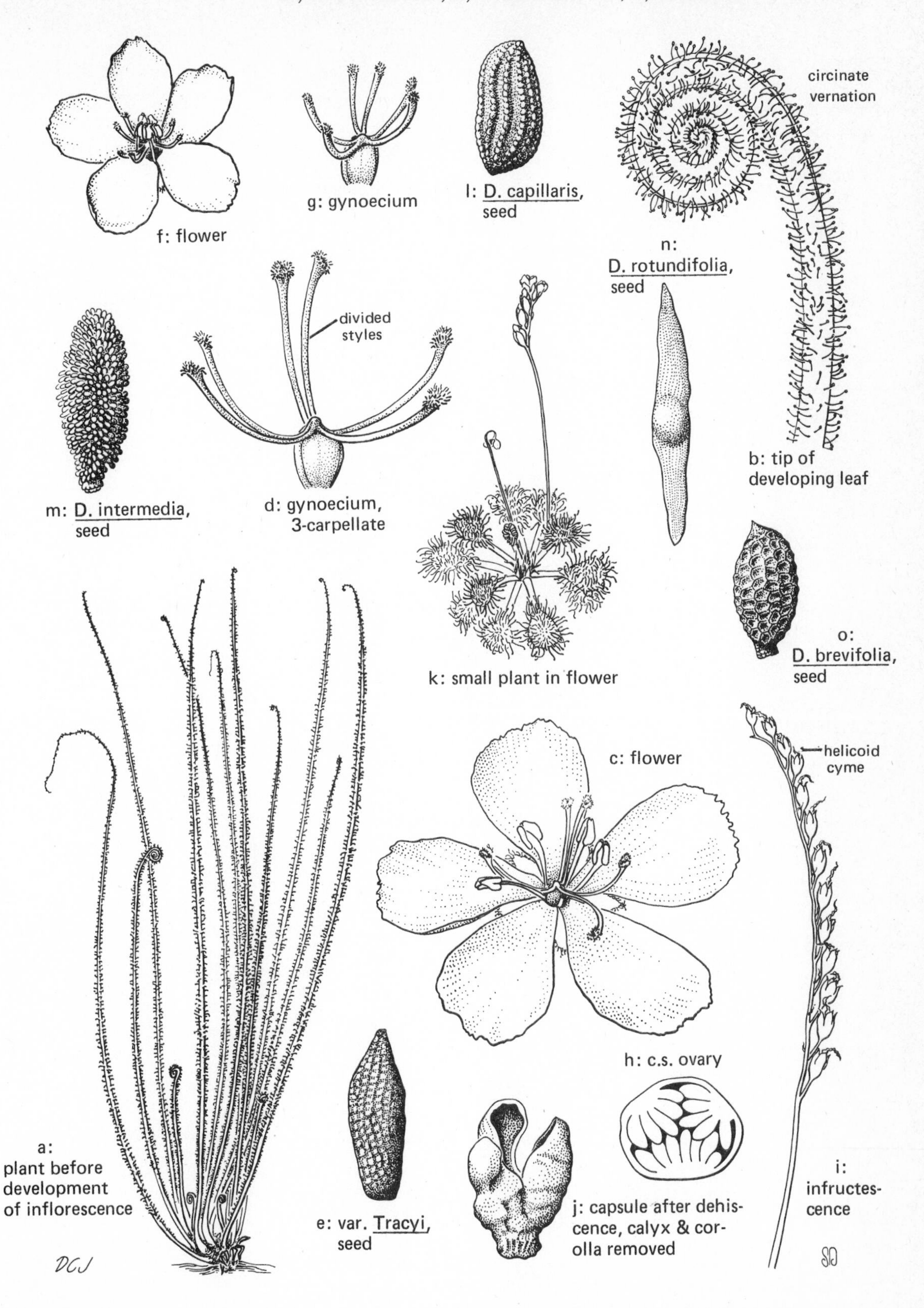

f: flower

g: gynoecium

l: D. capillaris, seed

circinate vernation

n: D. rotundifolia, seed

divided styles

b: tip of developing leaf

m: D. intermedia, seed

d: gynoecium, 3-carpellate

k: small plant in flower

o: D. brevifolia, seed

c: flower

helicoid cyme

a: plant before development of inflorescence

e: var. Tracyi, seed

h: c.s. ovary

j: capsule after dehiscence, calyx & corolla removed

i: infructescence

CRASSULACEAE: Sedum. a-d, S. pulchellum; e-j, S. pusillum; k, S. glaucophyllum; l, S. ternatum; m, S. telephioides

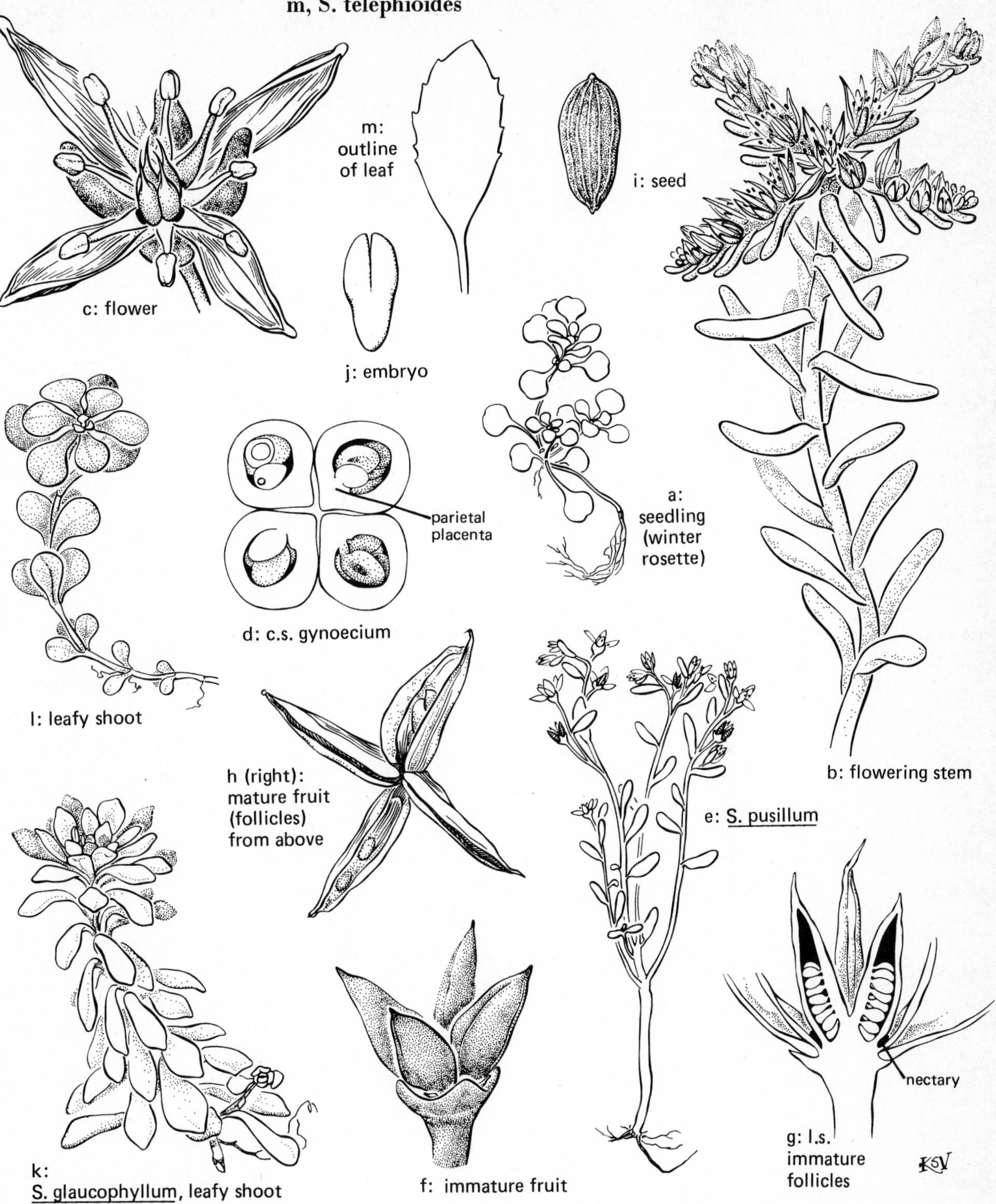

c: flower

m: outline of leaf

j: embryo

i: seed

d: c.s. gynoecium

parietal placenta

a: seedling (winter rosette)

l: leafy shoot

h (right): mature fruit (follicles) from above

b: flowering stem

e: S. pusillum

k: S. glaucophyllum, leafy shoot

f: immature fruit

g: l.s. immature follicles

nectary

SAXIFRAGACEAE: Tiarella. a-f, T. wherryi; g, T. cordifolia

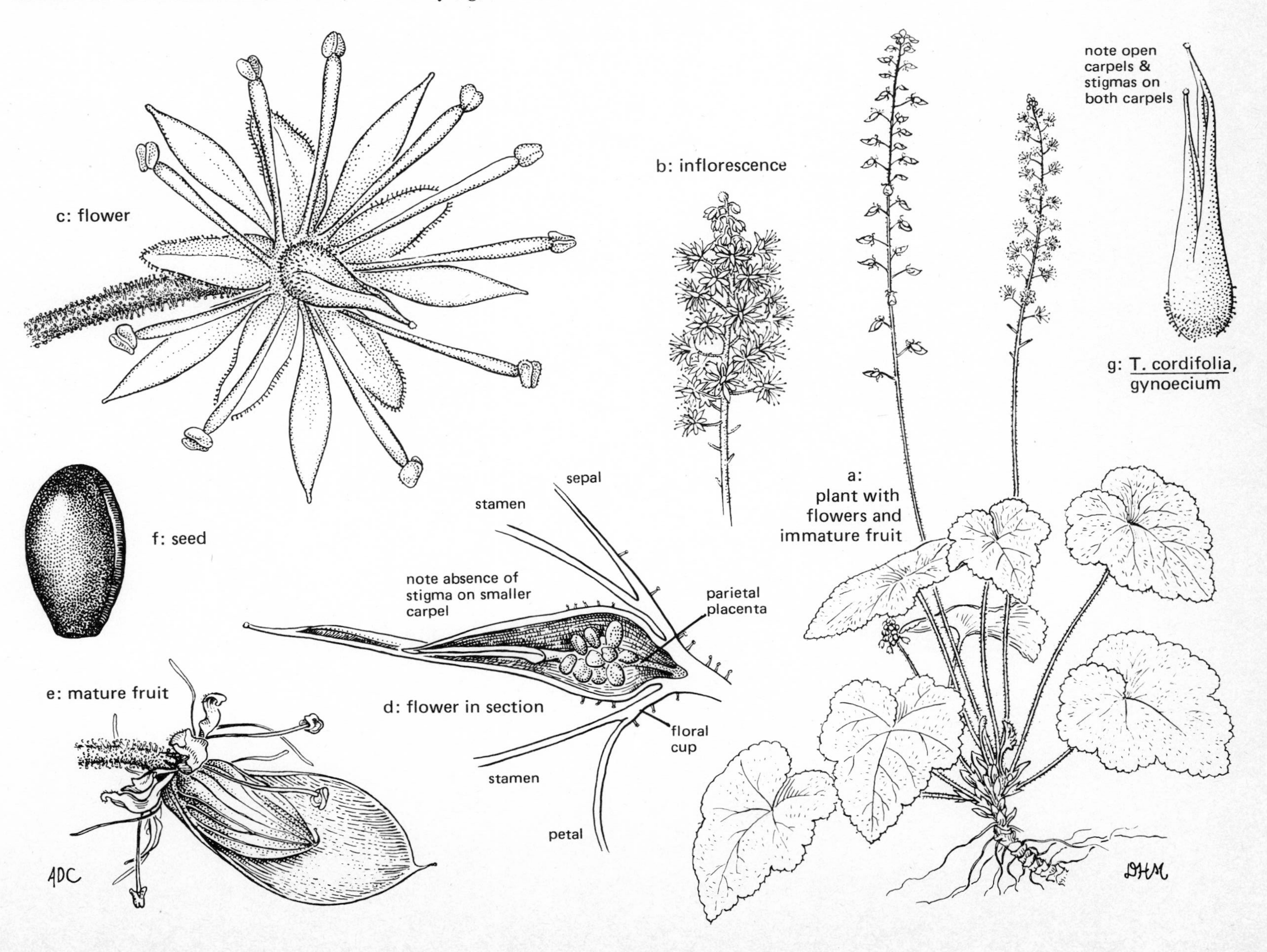

g: T. cordifolia, gynoecium

note open carpels & stigmas on both carpels

a: plant with flowers and immature fruit

b: inflorescence

c: flower

parietal placenta

sepal

floral cup

stamen

note absence of stigma on smaller carpel

petal

stamen

d: flower in section

f: seed

e: mature fruit

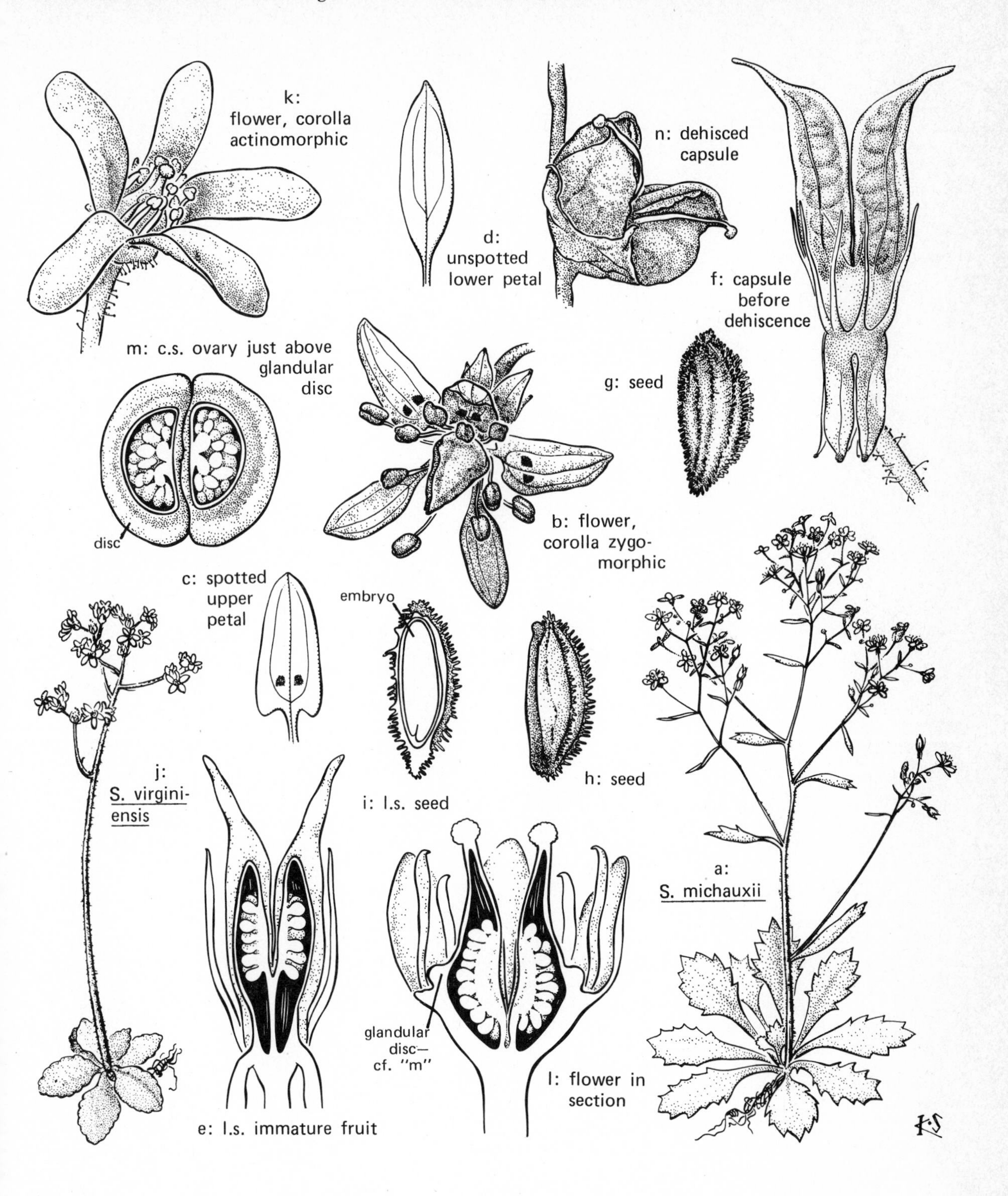

k: flower, corolla actinomorphic

d: unspotted lower petal

n: dehisced capsule

f: capsule before dehiscence

m: c.s. ovary just above glandular disc

g: seed

disc

b: flower, corolla zygo-morphic

c: spotted upper petal

embryo

j: S. virgini-ensis

i: l.s. seed

h: seed

a: S. michauxii

glandular disc— cf. "m"

l: flower in section

e: l.s. immature fruit

K·S

48

SAXIFRAGACEAE: **Philadelphus. a-d, P. inodorus; e-j, P. hirsutus**

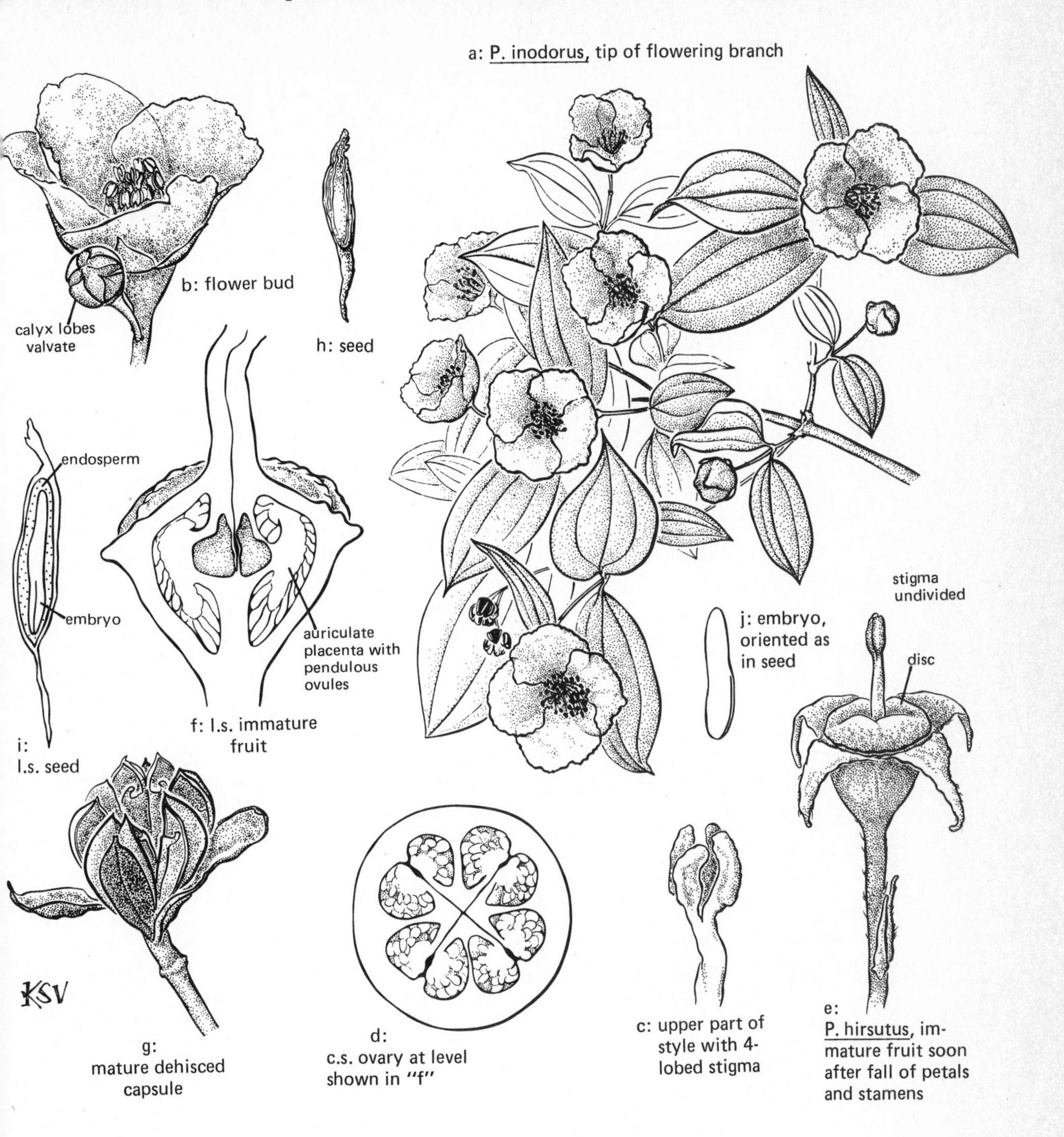

a: <u>P. inodorus,</u> tip of flowering branch

b: flower bud

calyx lobes
valvate

h: seed

endosperm

embryo

auriculate
placenta with
pendulous
ovules

i:
l.s. seed

f: l.s. immature
fruit

stigma
undivided

j: embryo,
oriented as
in seed

disc

KSV

g:
mature dehisced
capsule

d:
c.s. ovary at level
shown in "f"

c: upper part of
style with 4-
lobed stigma

e:
<u>P. hirsutus,</u> im-
mature fruit soon
after fall of petals
and stamens

HAMAMELIDACEAE: Hamamelis. a-k, H. vernalis; l, H. virginiana

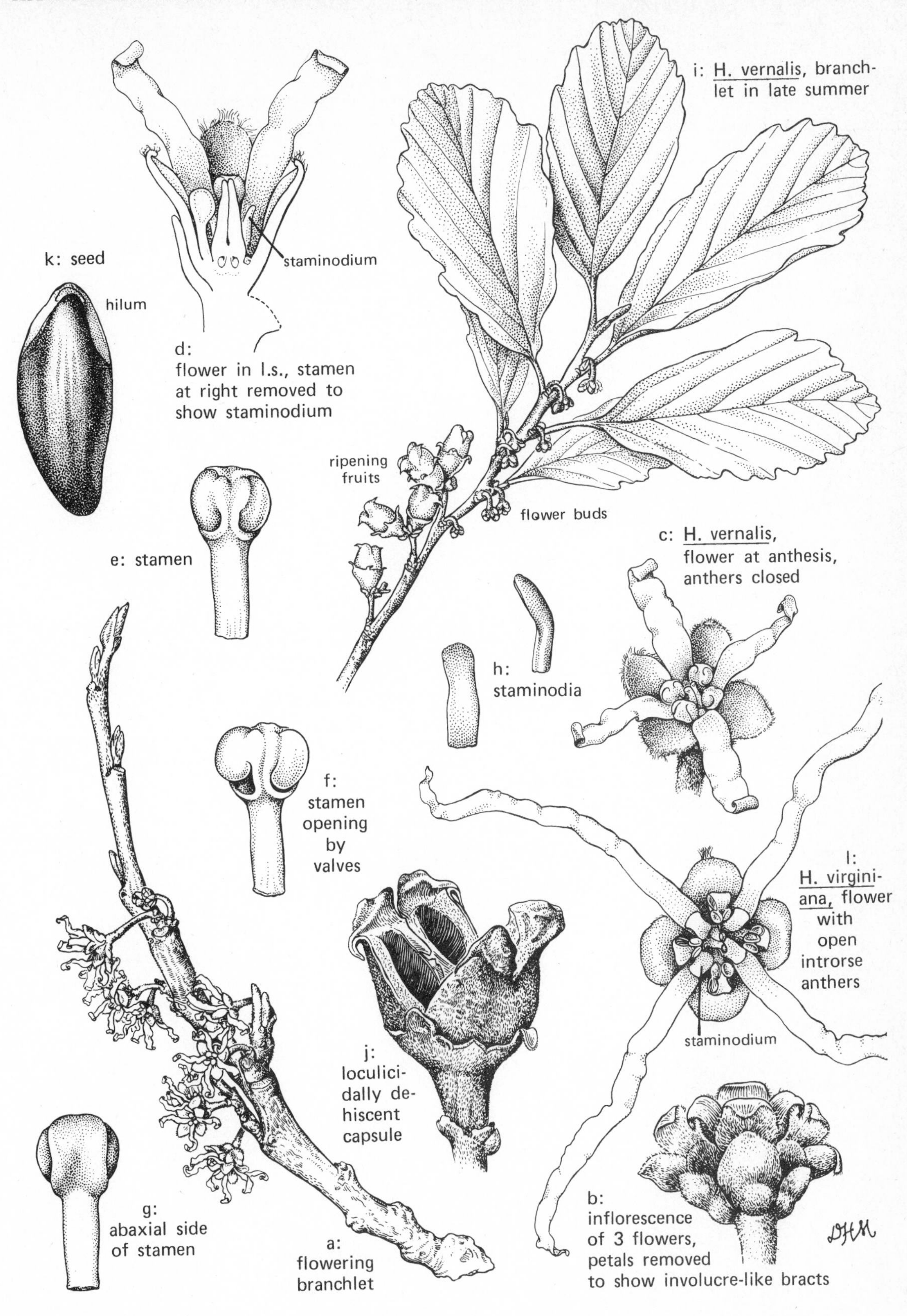

k: seed

hilum

staminodium

d:
flower in l.s., stamen
at right removed to
show staminodium

i: <u>H. vernalis</u>, branchlet in late summer

e: stamen

ripening
fruits

flower buds

h:
staminodia

c: <u>H. vernalis</u>,
flower at anthesis,
anthers closed

f:
stamen
opening
by
valves

l:
<u>H. virgini-
ana</u>, flower
with
open
introrse
anthers

staminodium

j:
loculici-
dally de-
hiscent
capsule

g:
abaxial side
of stamen

a:
flowering
branchlet

b:
inflorescence
of 3 flowers,
petals removed
to show involucre-like bracts

DHM

ROSACEAE subfamily SPIRAEOIDEAE: Physocarpus. a-e, P. opulifolius

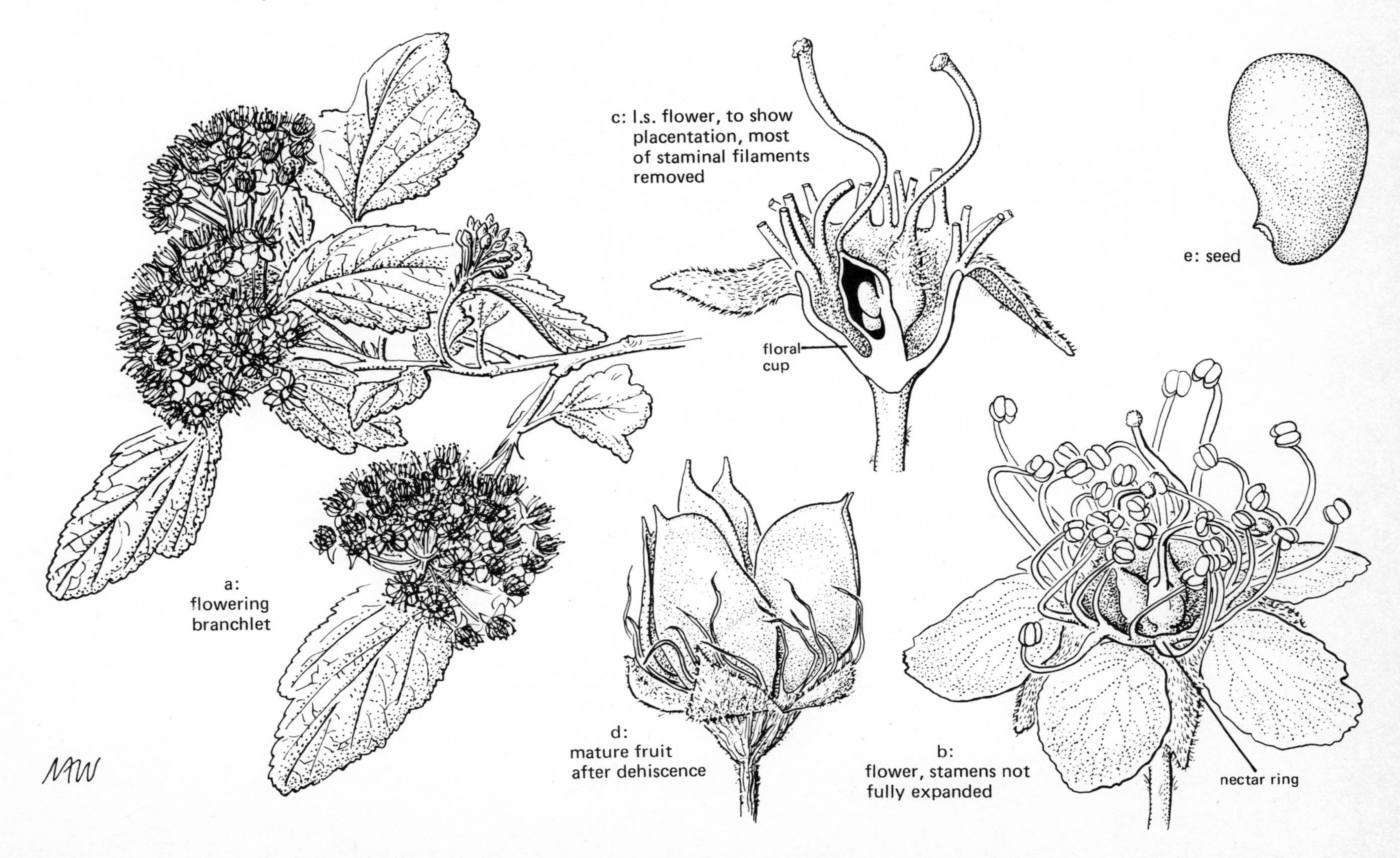

e: seed

nectar ring

b:
flower, stamens not
fully expanded

c: l.s. flower, to show
placentation, most
of staminal filaments
removed

floral
cup

d:
mature fruit
after dehiscence

a:
flowering
branchlet

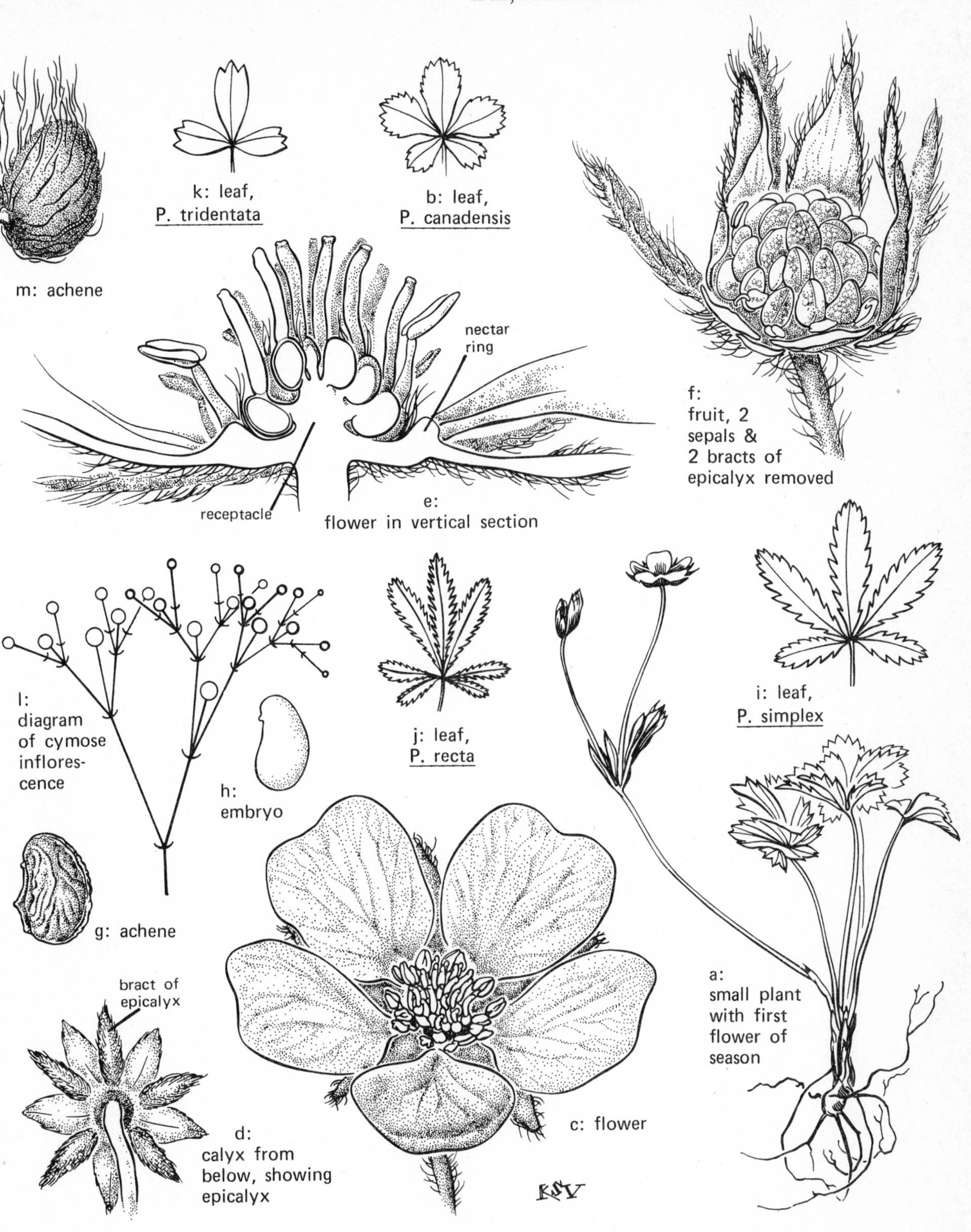

m: achene

k: leaf,
P. tridentata

b: leaf,
P. canadensis

nectar
ring

receptacle

e:
flower in vertical section

f:
fruit, 2
sepals &
2 bracts of
epicalyx removed

l:
diagram
of cymose
inflores-
cence

h:
embryo

j: leaf,
P. recta

i: leaf,
P. simplex

g: achene

bract of
epicalyx

d:
calyx from
below, showing
epicalyx

c: flower

a:
small plant
with first
flower of
season

KSV

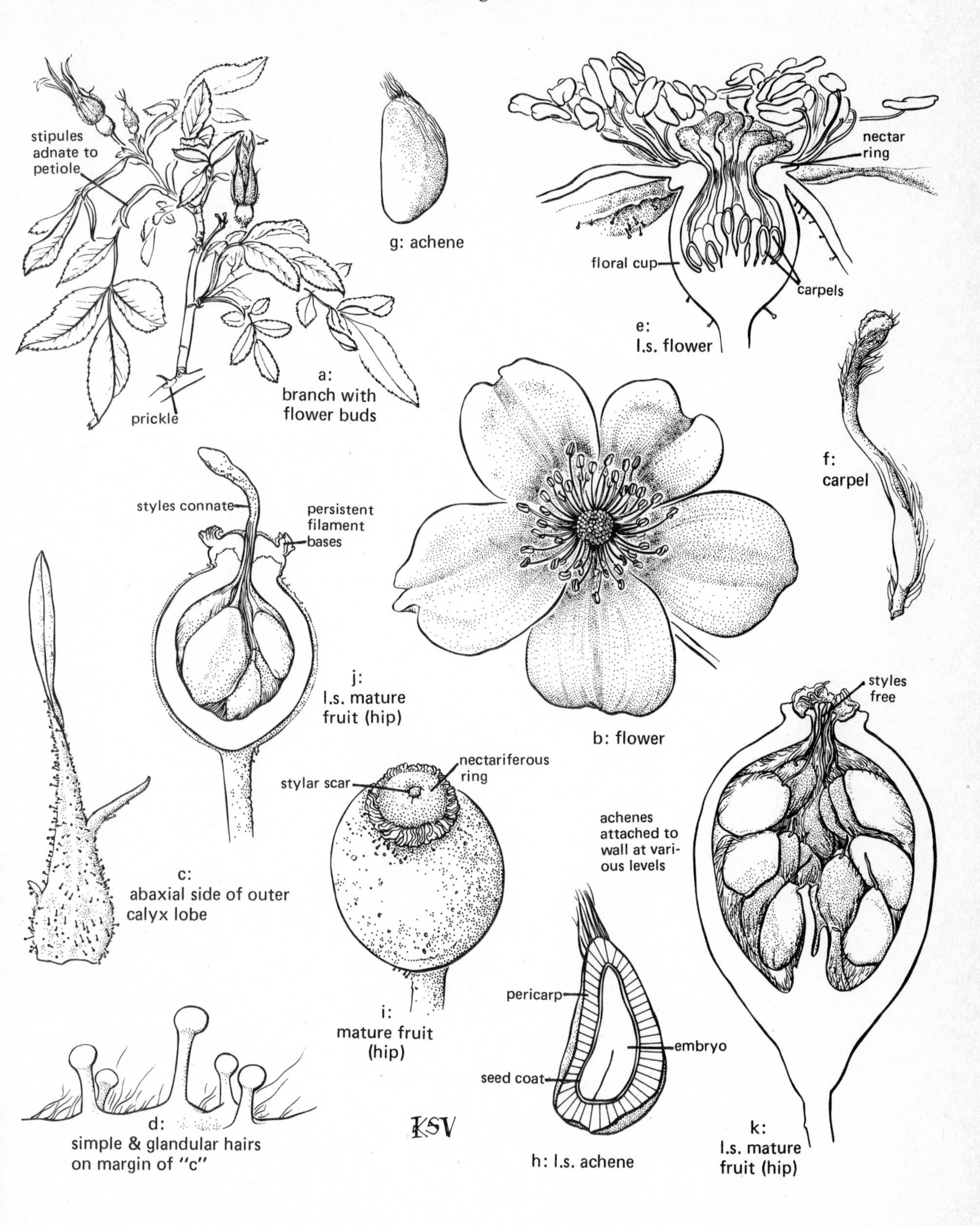

a: branch with flower buds

stipules adnate to petiole

prickle

g: achene

e: l.s. flower

nectar ring

floral cup

carpels

f: carpel

styles connate

persistent filament bases

j: l.s. mature fruit (hip)

b: flower

styles free

achenes attached to wall at various levels

stylar scar

nectariferous ring

c: abaxial side of outer calyx lobe

i: mature fruit (hip)

pericarp

embryo

seed coat

d: simple & glandular hairs on margin of "c"

KSV

h: l.s. achene

k: l.s. mature fruit (hip)

ROSACEAE subfamily AMYGDALOIDEAE: Prunus. a, P. virginiana; b-i, P. serotina; j-l, P. pensylvanica; m, P. caroliniana

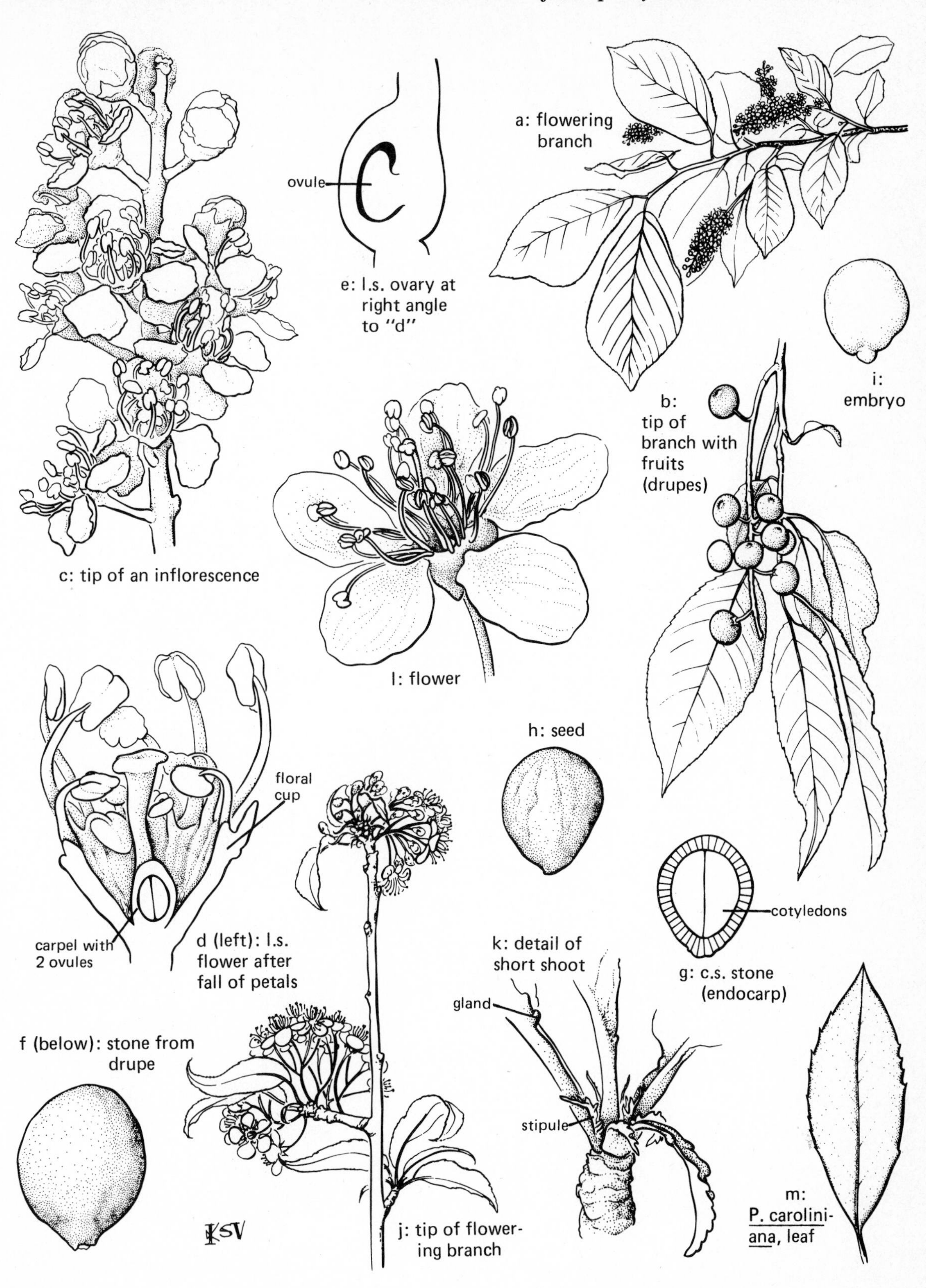

a: flowering branch

ovule

e: l.s. ovary at right angle to "d"

b: tip of branch with fruits (drupes)

i: embryo

c: tip of an inflorescence

l: flower

h: seed

floral cup

carpel with 2 ovules

d (left): l.s. flower after fall of petals

k: detail of short shoot

cotyledons

g: c.s. stone (endocarp)

f (below): stone from drupe

gland

stipule

j: tip of flowering branch

m: P. caroliniana, leaf

KSV

ROSACEAE subfamily MALOIDEAE: Pyrus subgenus Aronia. a–j, P. arbutifolia

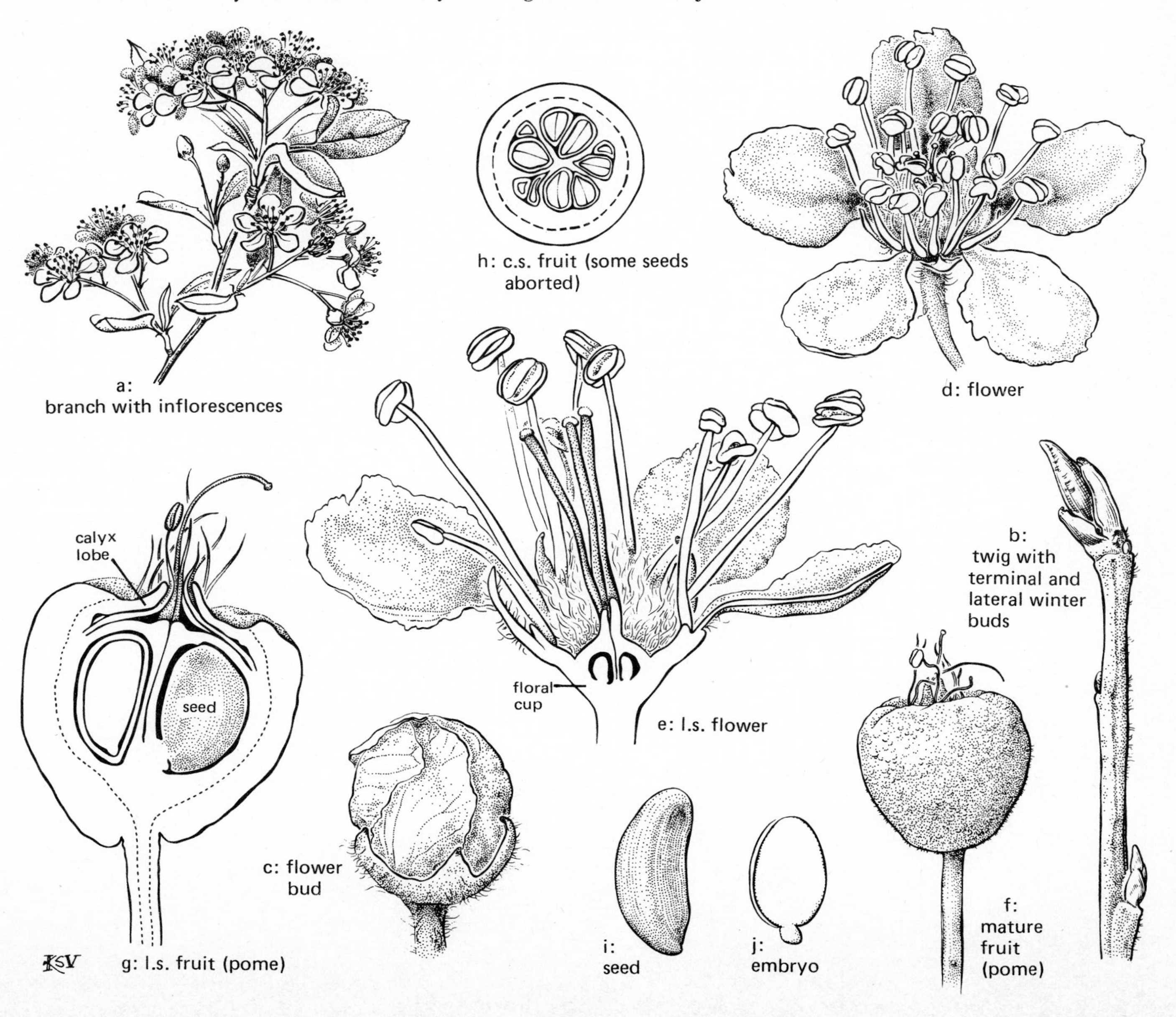

a: branch with inflorescences

b: twig with terminal and lateral winter buds

c: flower bud

d: flower

e: l.s. flower

floral cup

f: mature fruit (pome)

g: l.s. fruit (pome)

seed

calyx lobe

h: c.s. fruit (some seeds aborted)

i: seed

j: embryo

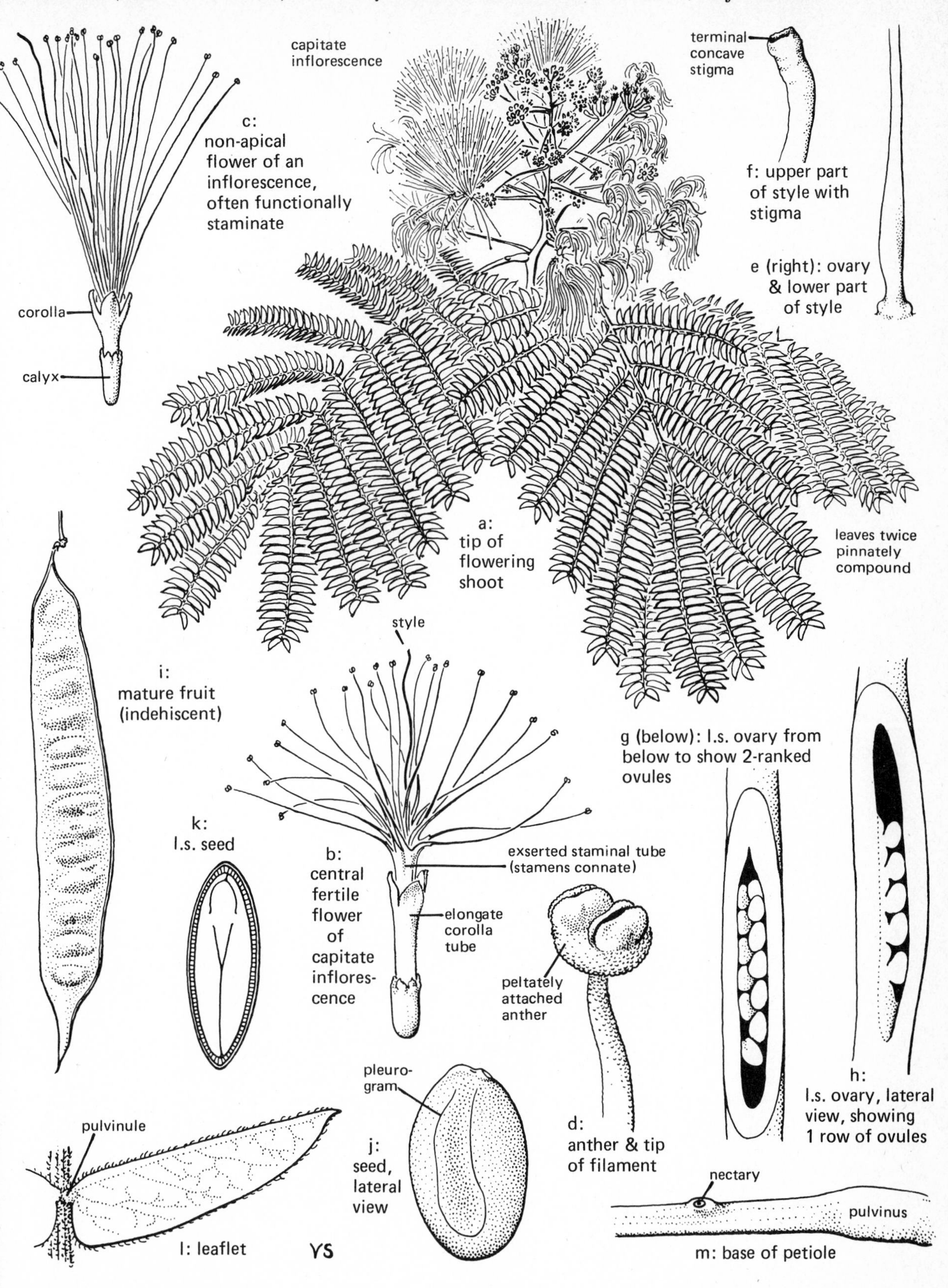

c: non-apical flower of an inflorescence, often functionally staminate

corolla

calyx

capitate inflorescence

terminal concave stigma

f: upper part of style with stigma

e (right): ovary & lower part of style

a: tip of flowering shoot

leaves twice pinnately compound

style

i: mature fruit (indehiscent)

g (below): l.s. ovary from below to show 2-ranked ovules

k: l.s. seed

b: central fertile flower of capitate inflorescence

exserted staminal tube (stamens connate)

elongate corolla tube

peltately attached anther

h: l.s. ovary, lateral view, showing 1 row of ovules

pleurogram

j: seed, lateral view

d: anther & tip of filament

pulvinule

l: leaflet   YS

nectary

pulvinus

m: base of petiole

56

FABACEAE (LEGUMINOSAE) subfamily CAESALPINOIDEAE: Cassia. a-j, C. bahamensis; k, l, C. tora; m-o, C. fasciculata

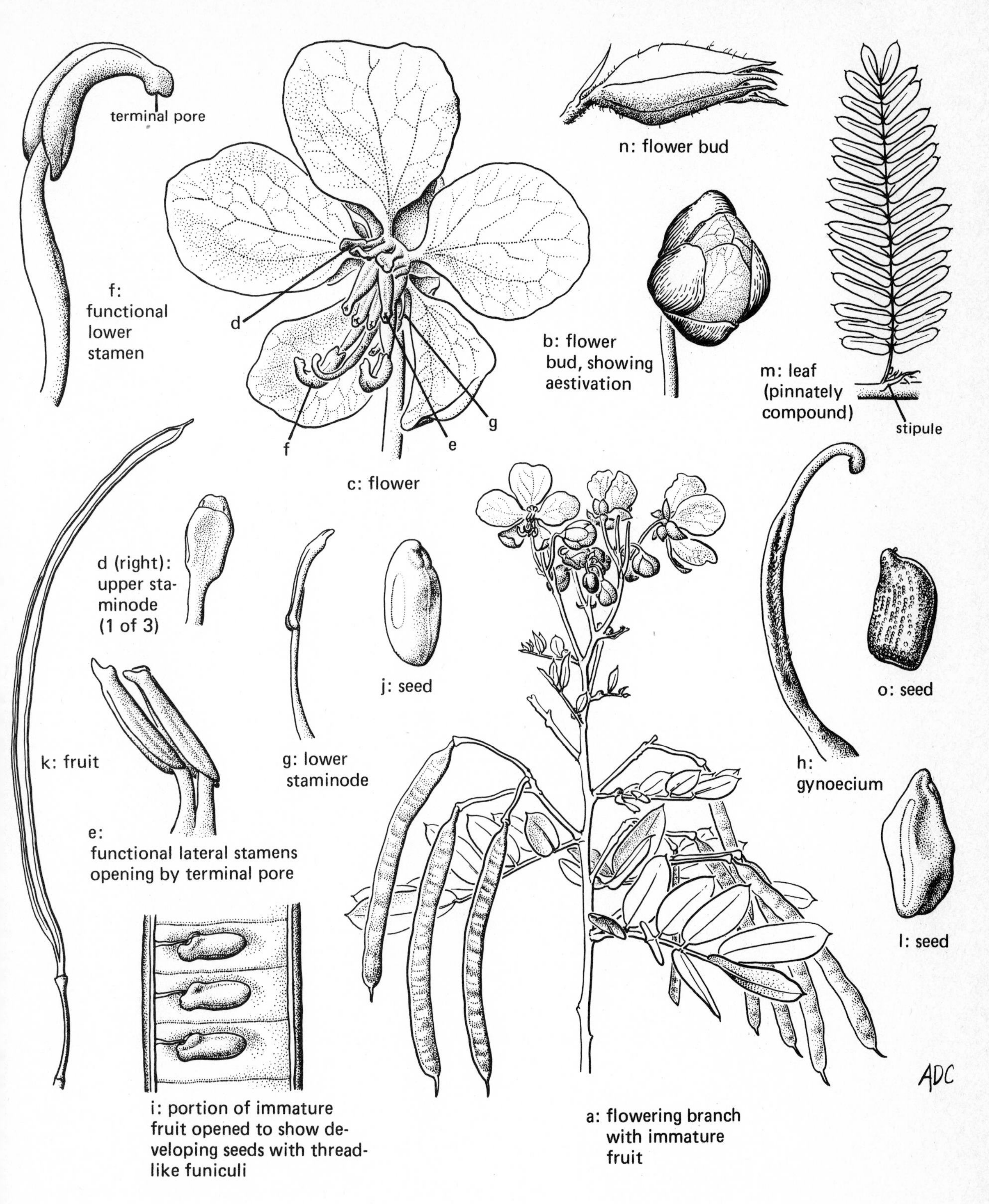

terminal pore

f: functional lower stamen

d

f

e

g

c: flower

n: flower bud

b: flower bud, showing aestivation

m: leaf (pinnately compound)

stipule

d (right): upper staminode (1 of 3)

j: seed

o: seed

k: fruit

g: lower staminode

h: gynoecium

l: seed

e: functional lateral stamens opening by terminal pore

i: portion of immature fruit opened to show developing seeds with thread-like funiculi

a: flowering branch with immature fruit

ADC

FABACEAE (LEGUMINOSAE) subfamily CAESALPINIOIDEAE: Cercis. a-j, C. canadensis

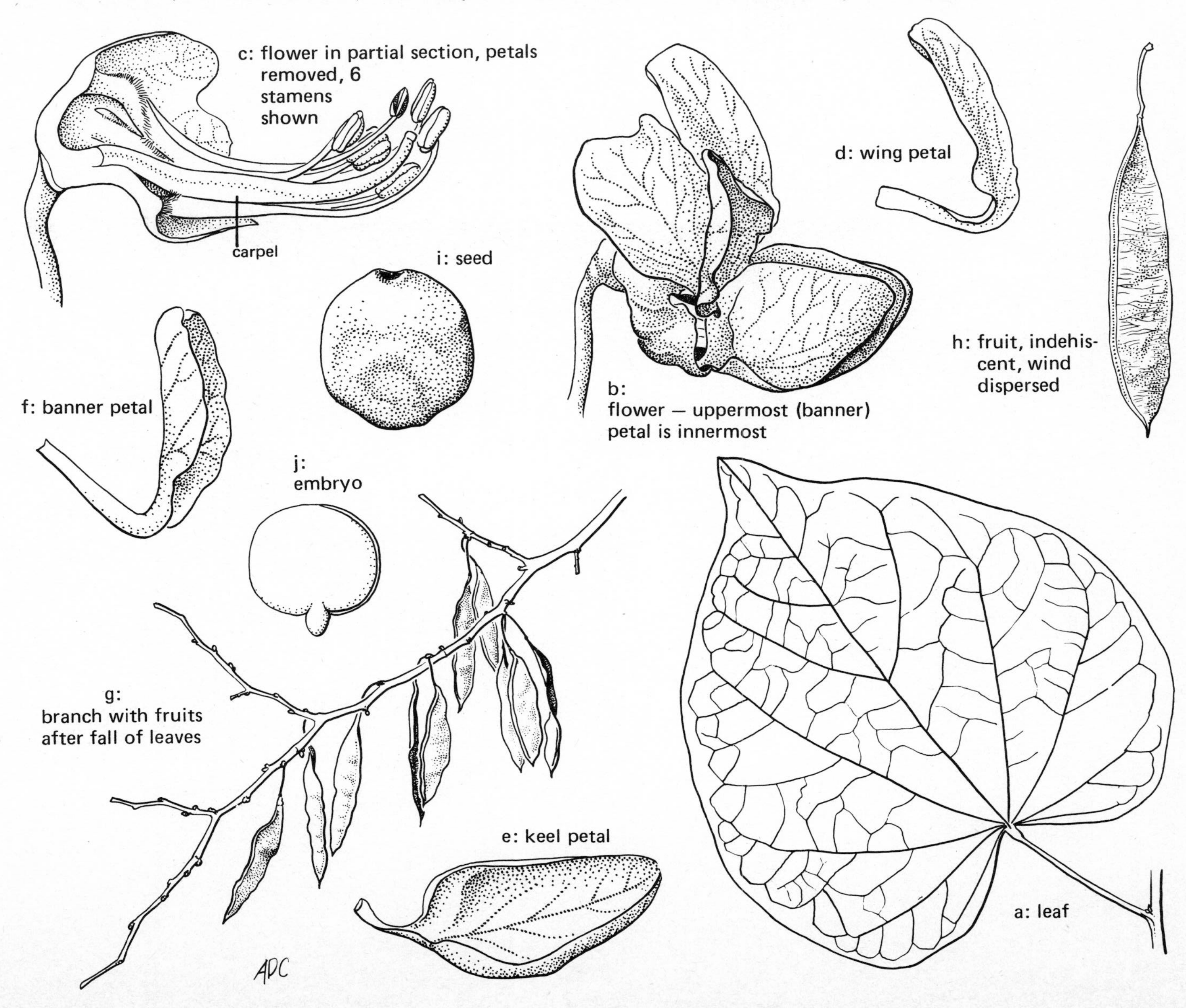

a: leaf

b: flower — uppermost (banner) petal is innermost

c: flower in partial section, petals removed, 6 stamens shown

carpel

d: wing petal

e: keel petal

f: banner petal

g: branch with fruits after fall of leaves

h: fruit, indehiscent, wind dispersed

i: seed

j: embryo

APC

FABACEAE (LEGUMINOSAE) subfamily FABOIDEAE: Baptisia. a-m, B. australis; n, B. tinctoria; o, B. arachnifera

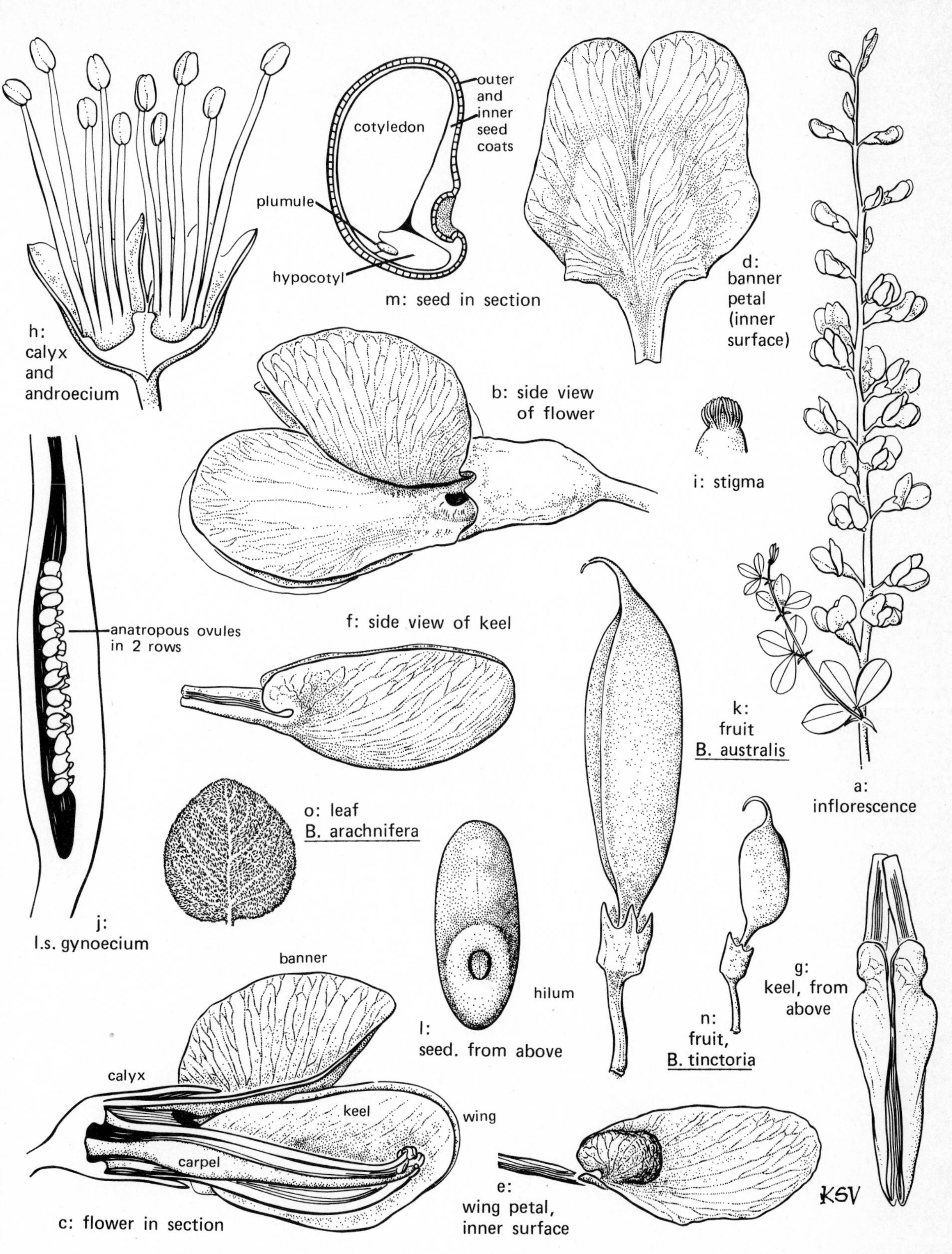

outer and inner seed coats

cotyledon

plumule

hypocotyl

m: seed in section

h: calyx and androecium

d: banner petal (inner surface)

b: side view of flower

i: stigma

anatropous ovules in 2 rows

f: side view of keel

k: fruit B. australis

a: inflorescence

o: leaf B. arachnifera

j: l.s. gynoecium

hilum

l: seed. from above

n: fruit, B. tinctoria

g: keel, from above

banner

calyx

keel

carpel

wing

c: flower in section

e: wing petal, inner surface

KSV

FABACEAE (LEGUMINOSAE) subfamily FABOIDEAE: Vicia. a-p, V. ludoviciana

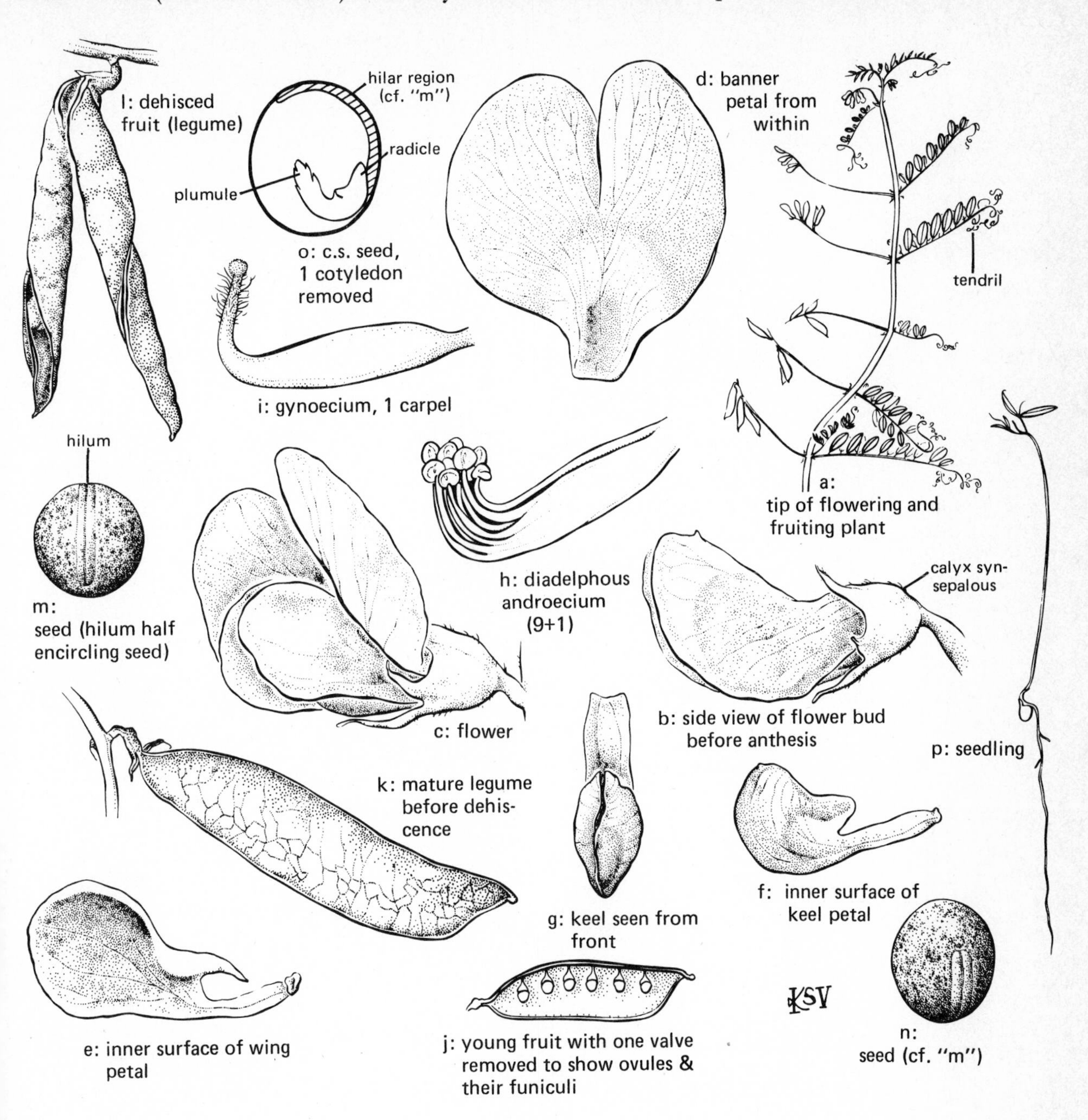

l: dehisced fruit (legume)

hilar region (cf. "m")

radicle

plumule

o: c.s. seed, 1 cotyledon removed

i: gynoecium, 1 carpel

d: banner petal from within

tendril

a: tip of flowering and fruiting plant

hilum

m: seed (hilum half encircling seed)

h: diadelphous androecium (9+1)

c: flower

calyx synsepalous

b: side view of flower bud before anthesis

p: seedling

k: mature legume before dehiscence

g: keel seen from front

f: inner surface of keel petal

e: inner surface of wing petal

j: young fruit with one valve removed to show ovules & their funiculi

KSV

n: seed (cf. "m")

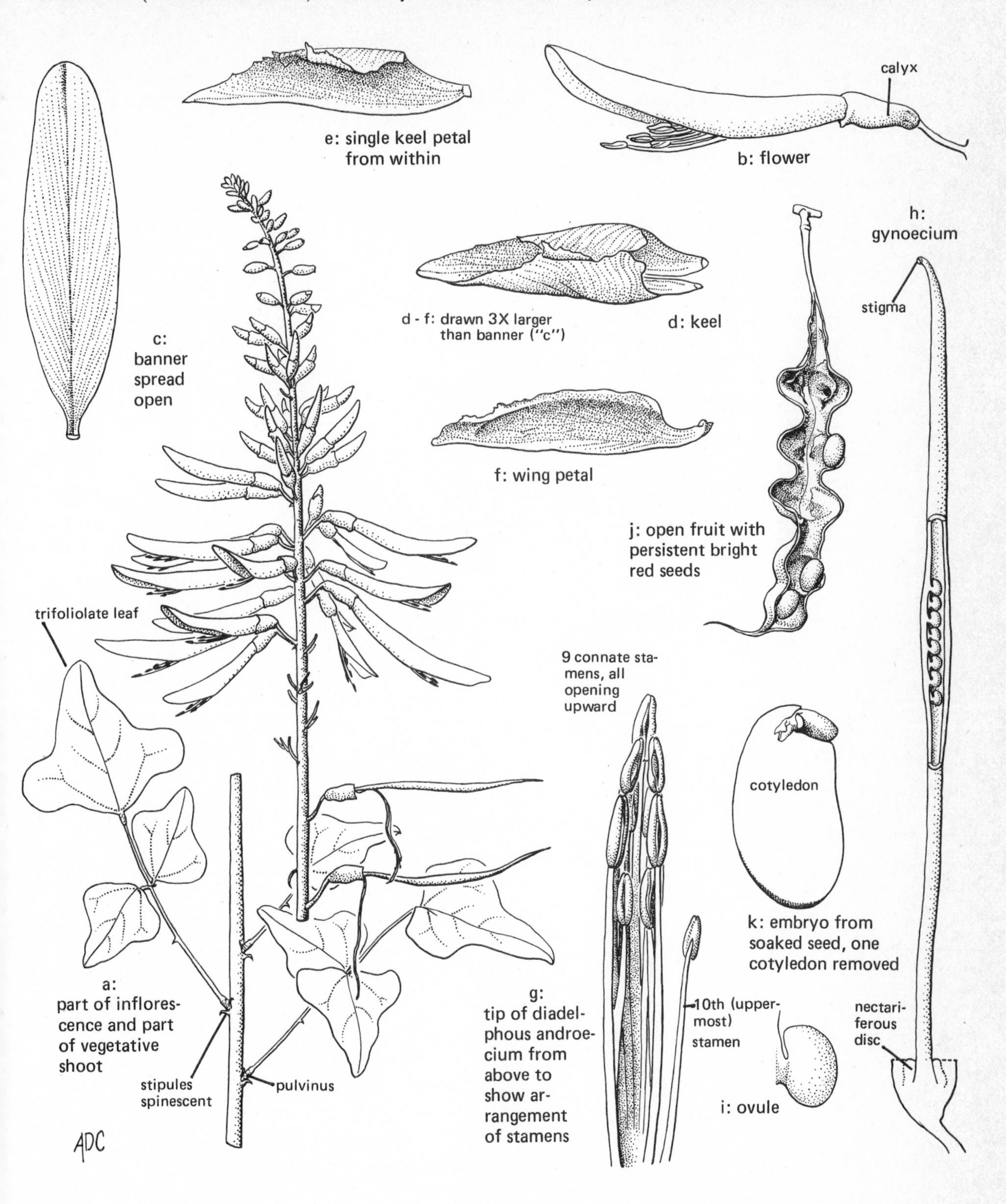

e: single keel petal from within

b: flower

calyx

h: gynoecium

stigma

c: banner spread open

d - f: drawn 3X larger than banner ("c")

d: keel

f: wing petal

j: open fruit with persistent bright red seeds

trifoliolate leaf

9 connate stamens, all opening upward

cotyledon

k: embryo from soaked seed, one cotyledon removed

a: part of inflorescence and part of vegetative shoot

stipules spinescent

pulvinus

g: tip of diadelphous androecium from above to show arrangement of stamens

10th (uppermost) stamen

i: ovule

nectariferous disc

ADC

GERANIACEAE: Geranium. a-j, G. maculatum; k-m, G. carolinianum

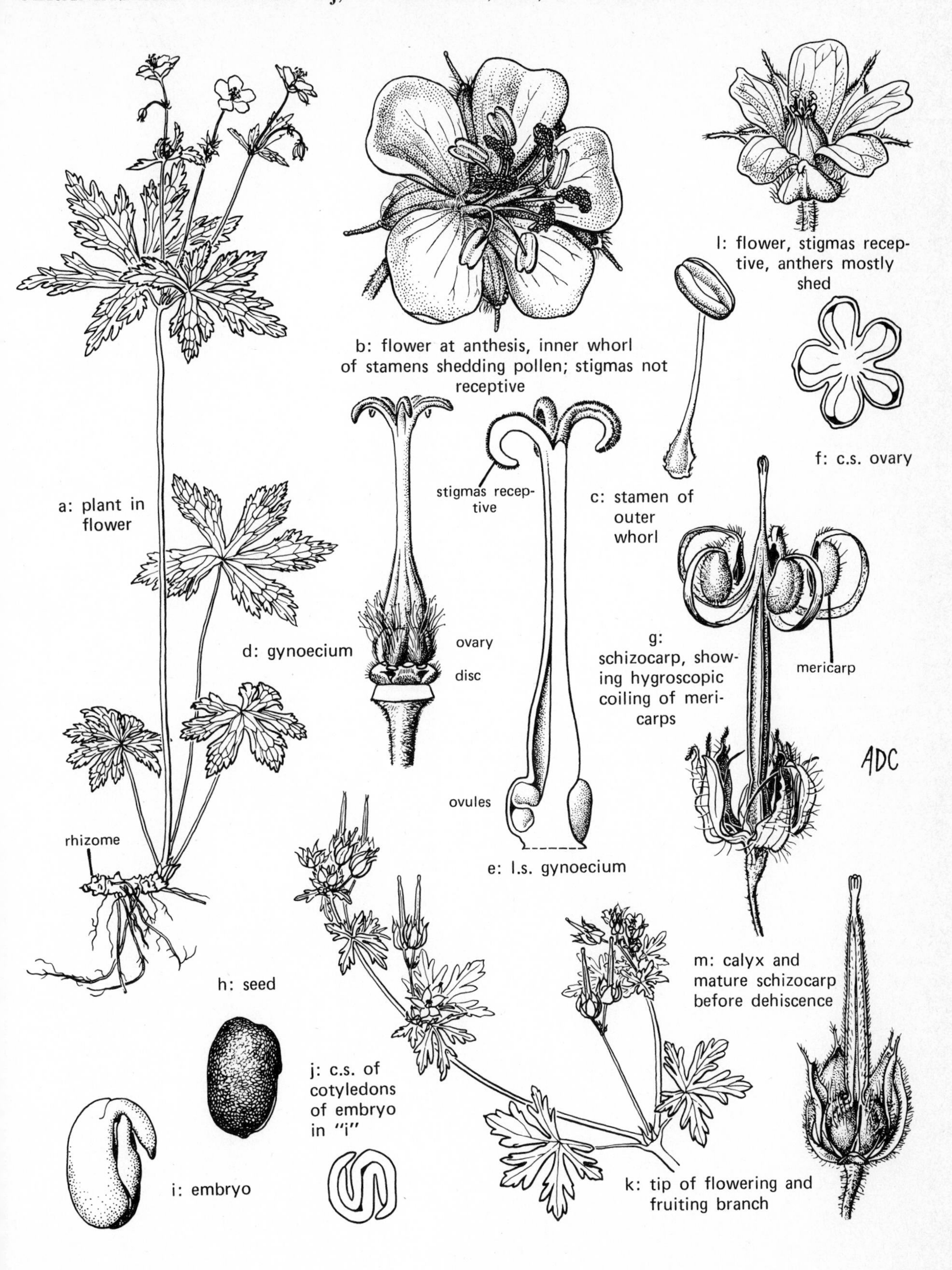

a: plant in flower

rhizome

b: flower at anthesis, inner whorl of stamens shedding pollen; stigmas not receptive

d: gynoecium

stigmas receptive

c: stamen of outer whorl

ovary

disc

ovules

e: l.s. gynoecium

l: flower, stigmas receptive, anthers mostly shed

f: c.s. ovary

g: schizocarp, showing hygroscopic coiling of mericarps

mericarp

ADC

h: seed

i: embryo

j: c.s. of cotyledons of embryo in "i"

k: tip of flowering and fruiting branch

m: calyx and mature schizocarp before dehiscence

62

RUTACEAE: Poncirus. a-j, P. trifoliata

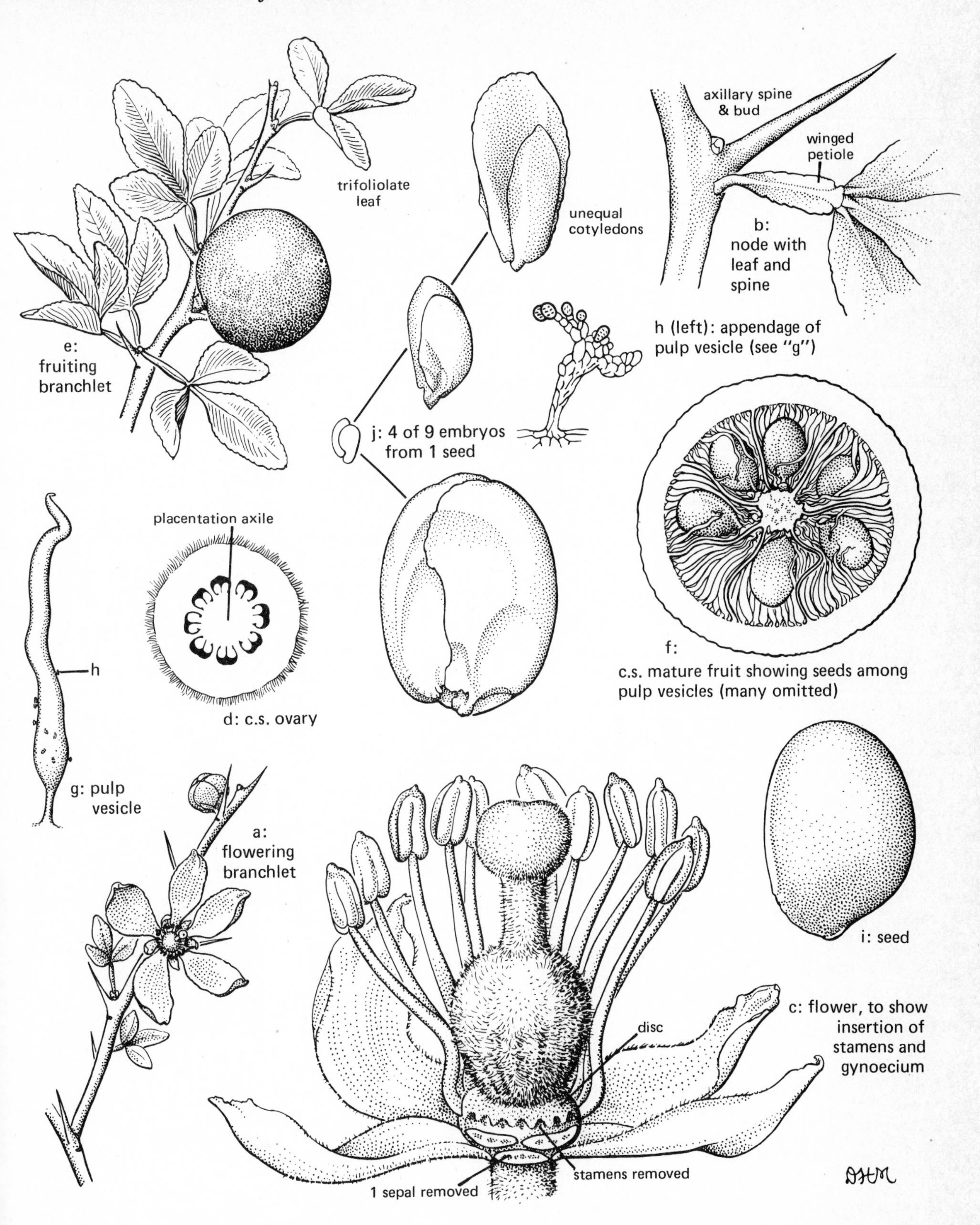

trifoliolate leaf

unequal cotyledons

axillary spine & bud

winged petiole

b: node with leaf and spine

h (left): appendage of pulp vesicle (see "g")

e: fruiting branchlet

j: 4 of 9 embryos from 1 seed

placentation axile

h

d: c.s. ovary

f: c.s. mature fruit showing seeds among pulp vesicles (many omitted)

g: pulp vesicle

a: flowering branchlet

i: seed

c: flower, to show insertion of stamens and gynoecium

disc

1 sepal removed

stamens removed

POLYGALACEAE: Polygala. a-k, P. paucifolia; l-o, P. grandiflora

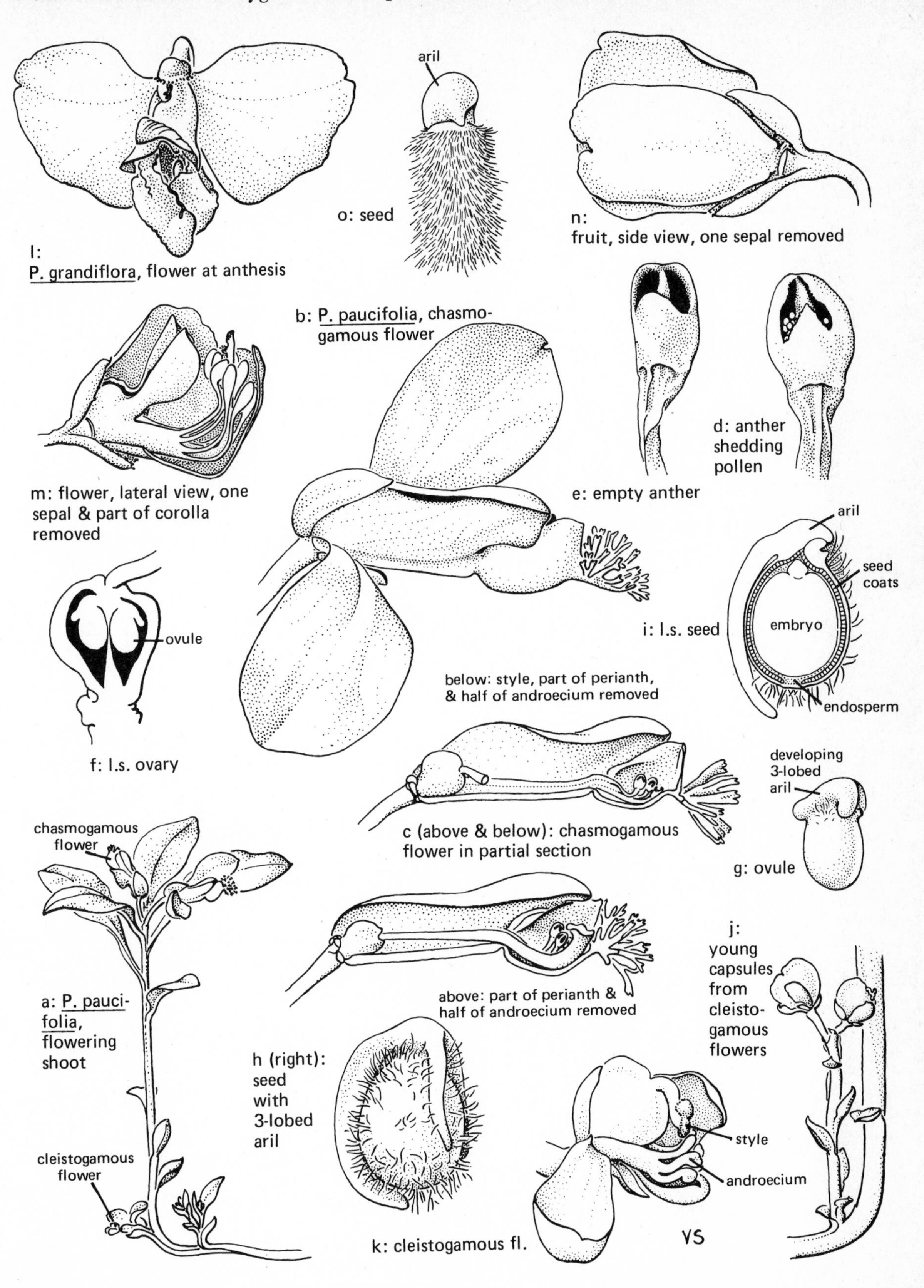

l:
P. grandiflora, flower at anthesis

o: seed

aril

n:
fruit, side view, one sepal removed

b: P. paucifolia, chasmo-
gamous flower

d: anther
shedding
pollen

e: empty anther

m: flower, lateral view, one
sepal & part of corolla
removed

i: l.s. seed

aril

seed
coats

embryo

endosperm

ovule

f: l.s. ovary

below: style, part of perianth,
& half of androecium removed

developing
3-lobed
aril

g: ovule

c (above & below): chasmogamous
flower in partial section

chasmogamous
flower

above: part of perianth &
half of androecium removed

j:
young
capsules
from
cleisto-
gamous
flowers

a: P. pauci-
folia,
flowering
shoot

h (right):
seed
with
3-lobed
aril

style

androecium

cleistogamous
flower

k: cleistogamous fl.

VS

EUPHORBIACEAE: Croton. a-b, C. linearis; c-m, C. alabamensis; n, o, C. glandulosus; p, C. punctatus; q, C. argyranthemus

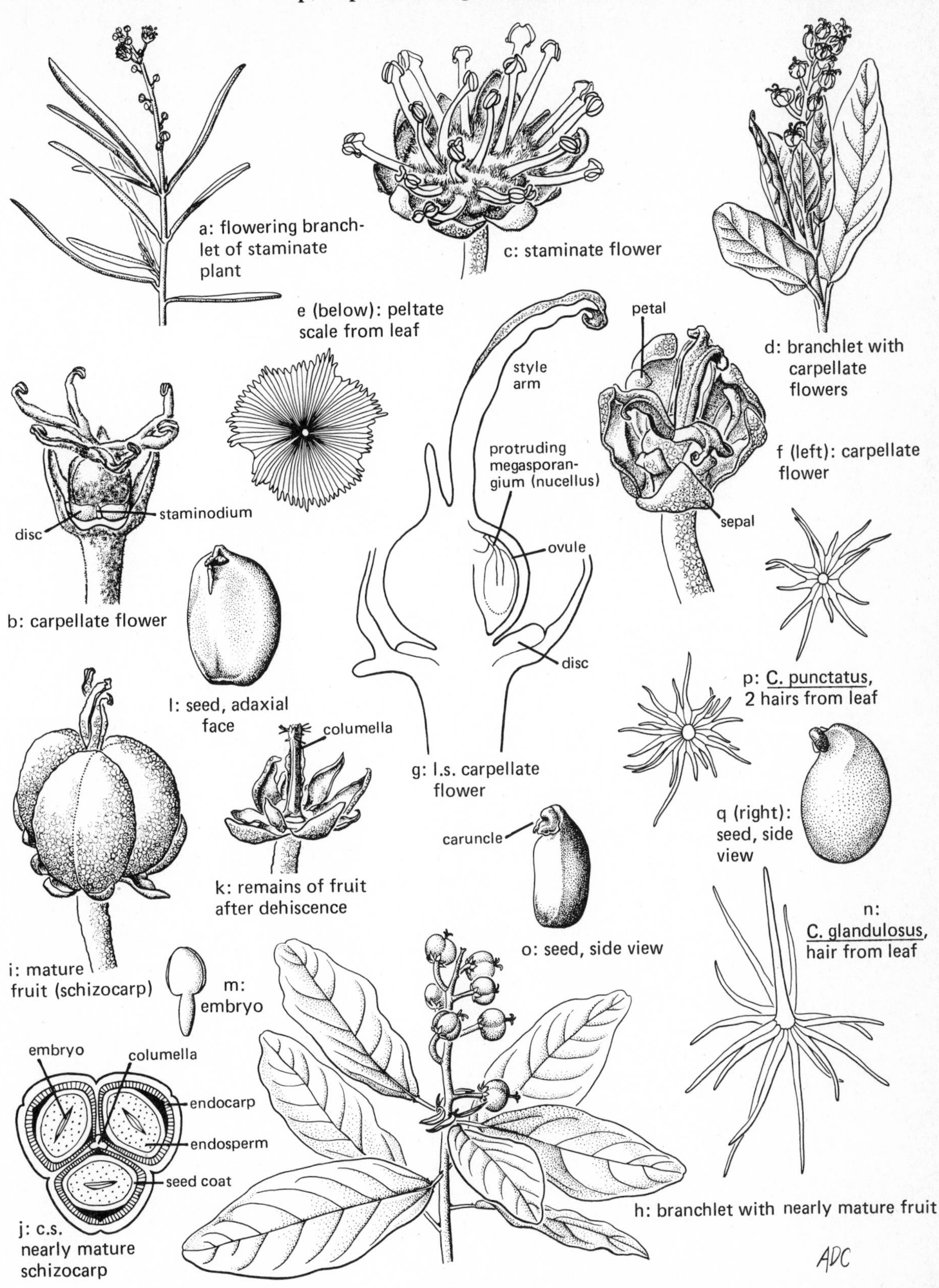

a: flowering branch-
let of staminate
plant

c: staminate flower

e (below): peltate
scale from leaf

petal

style
arm

protruding
megasporan-
gium (nucellus)

ovule

disc

d: branchlet with
carpellate
flowers

f (left): carpellate
flower

sepal

disc

staminodium

b: carpellate flower

l: seed, adaxial
face

columella

g: l.s. carpellate
flower

caruncle

k: remains of fruit
after dehiscence

p: C. punctatus,
2 hairs from leaf

q (right):
seed, side
view

n:
C. glandulosus,
hair from leaf

i: mature
fruit (schizocarp)

m:
embryo

o: seed, side view

embryo   columella

endocarp

endosperm

seed coat

j: c.s.
nearly mature
schizocarp

h: branchlet with nearly mature fruit

ADC

65

EUPHORBIACEAE: a-i, *Euphorbia corollata*; j, *E. inundata*; k, l, *E. commutata*;
m, *E. dentata*; n-p, *Chamaesyce maculata*

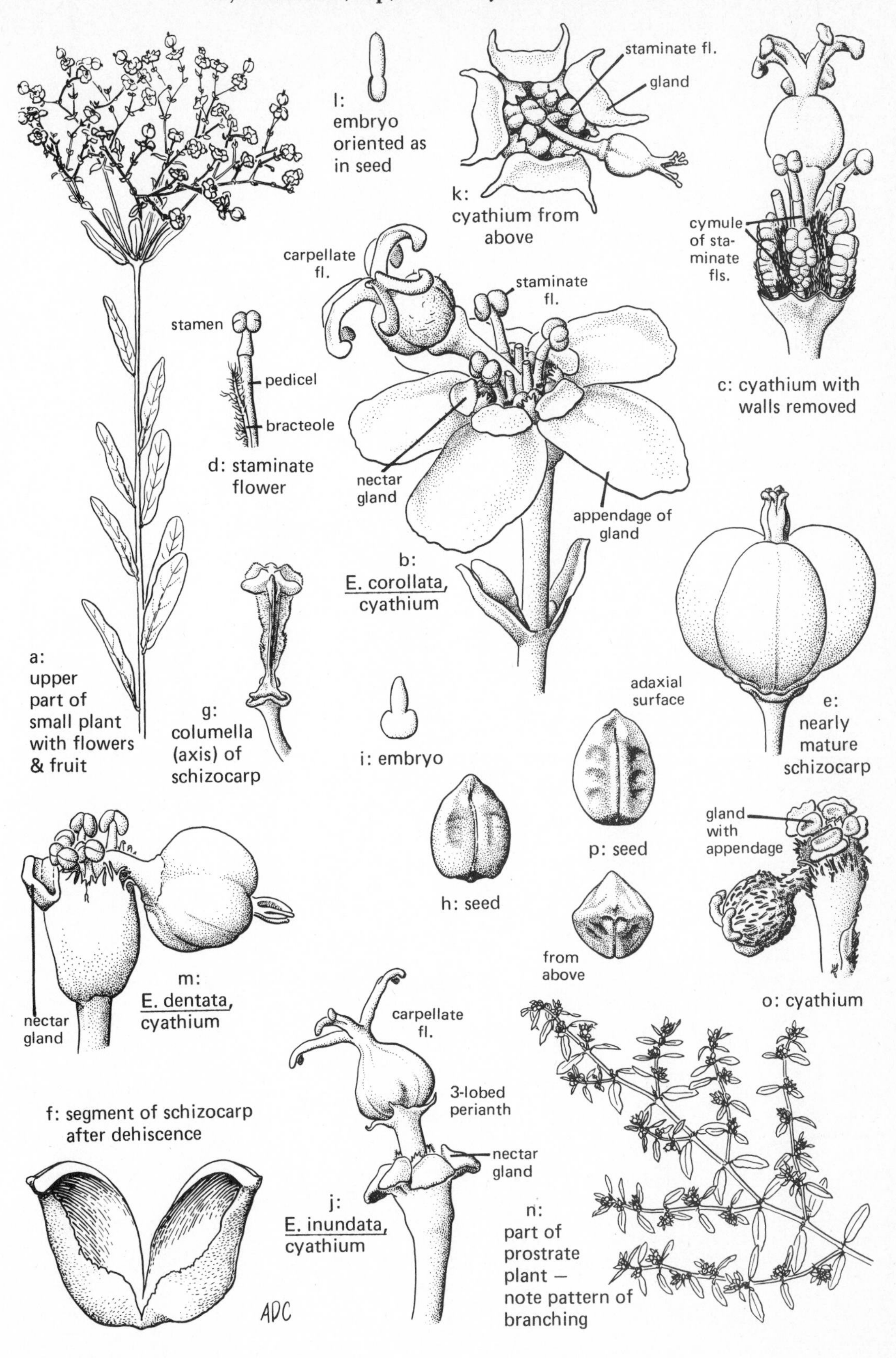

l: embryo oriented as in seed

k: cyathium from above

staminate fl.

gland

cymule of staminate fls.

c: cyathium with walls removed

carpellate fl.

staminate fl.

stamen

pedicel

bracteole

d: staminate flower

nectar gland

appendage of gland

b: E. corollata, cyathium

a: upper part of small plant with flowers & fruit

g: columella (axis) of schizocarp

i: embryo

h: seed

adaxial surface

p: seed

from above

e: nearly mature schizocarp

gland with appendage

o: cyathium

m: E. dentata, cyathium

nectar gland

f: segment of schizocarp after dehiscence

carpellate fl.

3-lobed perianth

nectar gland

j: E. inundata, cyathium

n: part of prostrate plant — note pattern of branching

ADC

AQUIFOLIACEAE: Ilex. a-i, I. glabra; j-n, I. opaca

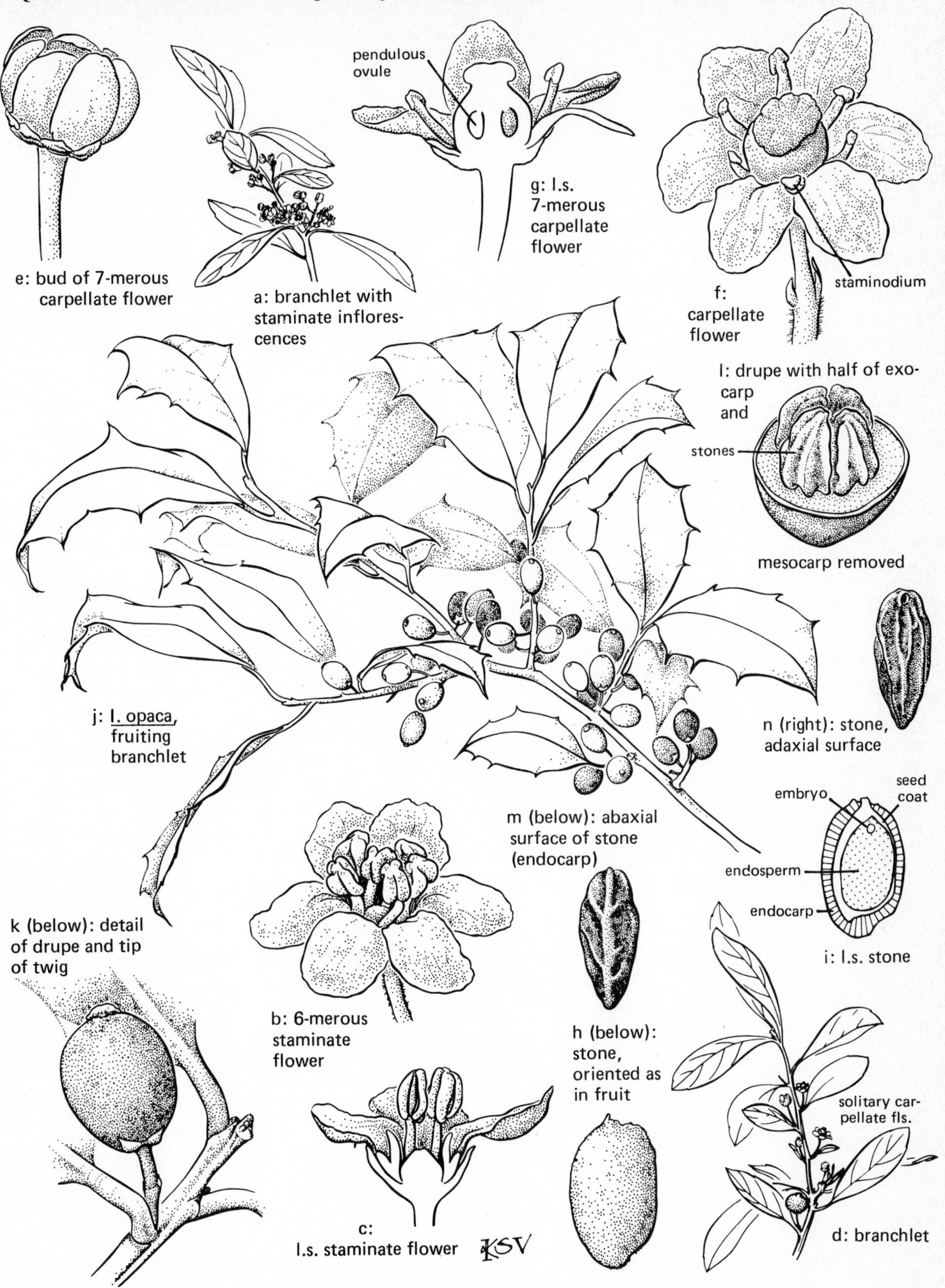

e: bud of 7-merous carpellate flower

a: branchlet with staminate inflorescences

pendulous ovule

g: l.s. 7-merous carpellate flower

f: carpellate flower

staminodium

l: drupe with half of exocarp and

stones

mesocarp removed

j: I. opaca, fruiting branchlet

n (right): stone, adaxial surface

m (below): abaxial surface of stone (endocarp)

embryo

seed coat

endosperm

endocarp

i: l.s. stone

k (below): detail of drupe and tip of twig

b: 6-merous staminate flower

h (below): stone, oriented as in fruit

c: l.s. staminate flower

KSV

solitary carpellate fls.

d: branchlet

ACERACEAE: Acer section Saccharina. a-h, A. saccharum subsp. saccharum; i-l, subsp. floridanum

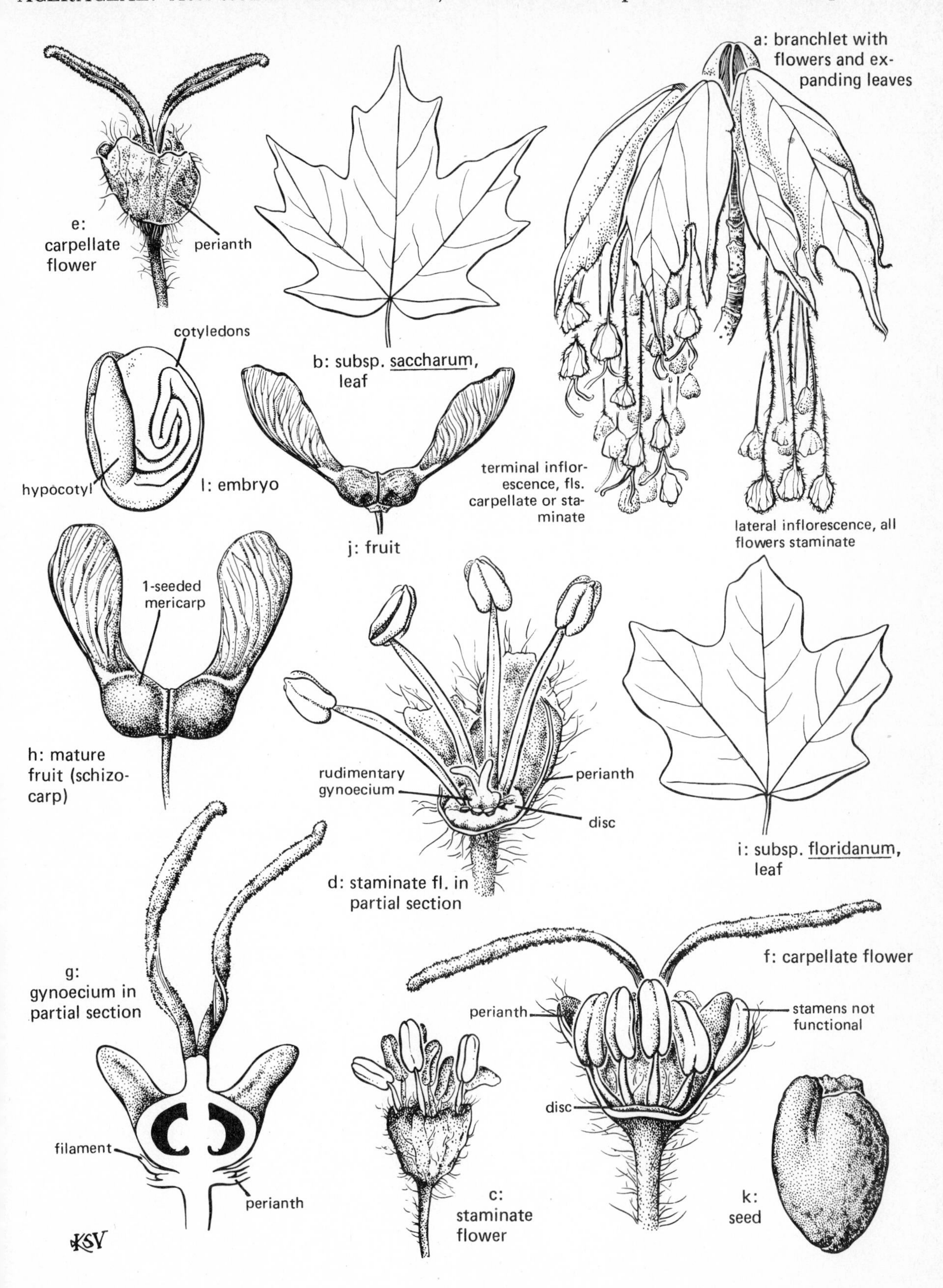

a: branchlet with flowers and expanding leaves

e: carpellate flower

perianth

cotyledons

hypocotyl   l: embryo

b: subsp. saccharum, leaf

j: fruit

terminal inflorescence, fls. carpellate or staminate

lateral inflorescence, all flowers staminate

1-seeded mericarp

h: mature fruit (schizocarp)

rudimentary gynoecium

perianth

disc

d: staminate fl. in partial section

i: subsp. floridanum, leaf

g: gynoecium in partial section

filament

perianth

c: staminate flower

perianth

disc

f: carpellate flower

stamens not functional

k: seed

KSV

68

MALVACEAE: Kosteletzkya. a-f, K. virginica var. altheifolia

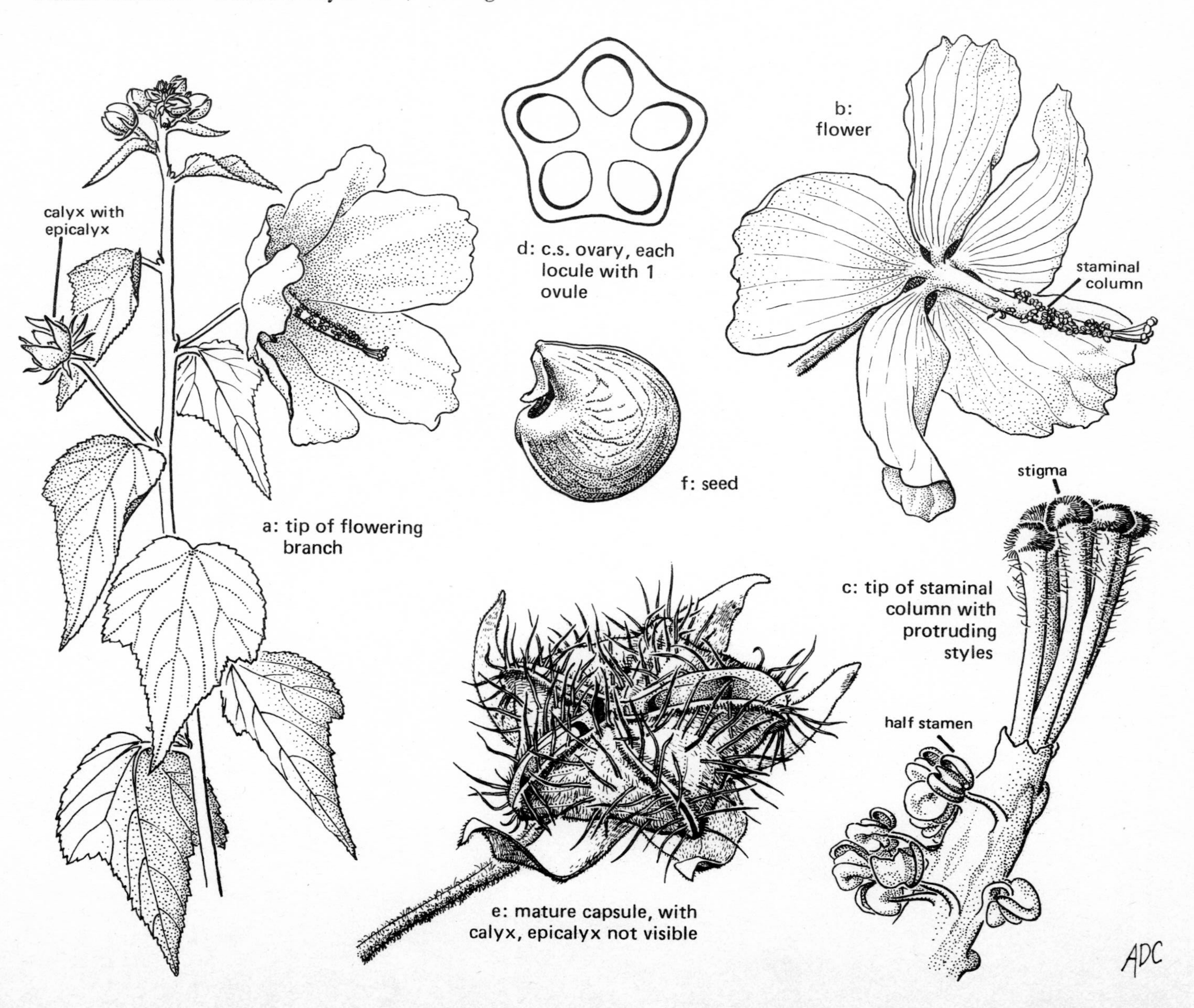

b: flower

staminal column

stigma

c: tip of staminal column with protruding styles

half stamen

d: c.s. ovary, each locule with 1 ovule

f: seed

e: mature capsule, with calyx, epicalyx not visible

a: tip of flowering branch

calyx with epicalyx

APC

MALVACEAE: Sida. a-i, S. acuta

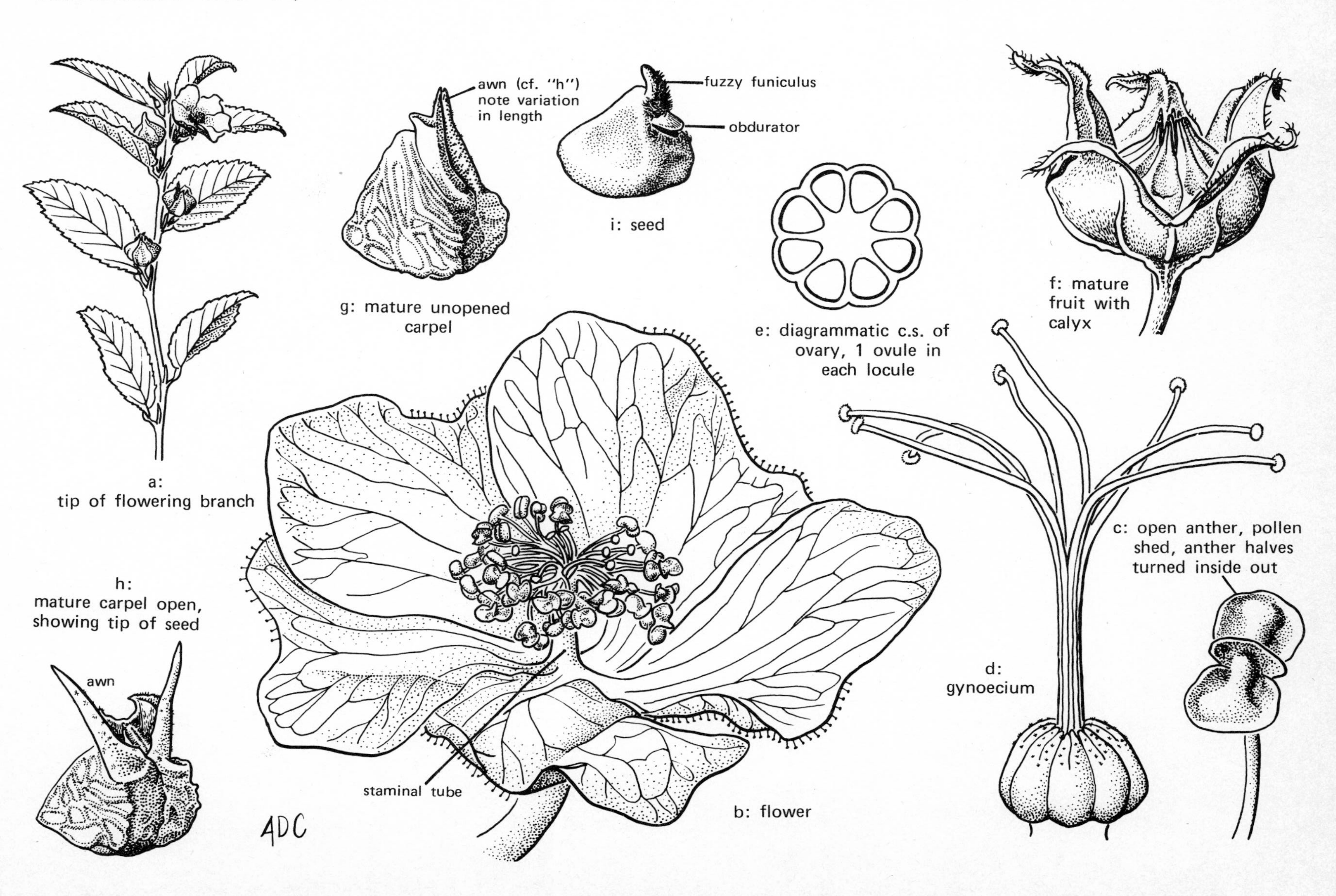

a: tip of flowering branch

b: flower

staminal tube

c: open anther, pollen shed, anther halves turned inside out

d: gynoecium

e: diagrammatic c.s. of ovary, 1 ovule in each locule

f: mature fruit with calyx

g: mature unopened carpel

awn (cf. "h") note variation in length

h: mature carpel open, showing tip of seed

awn

i: seed

fuzzy funiculus

obdurator

ADC

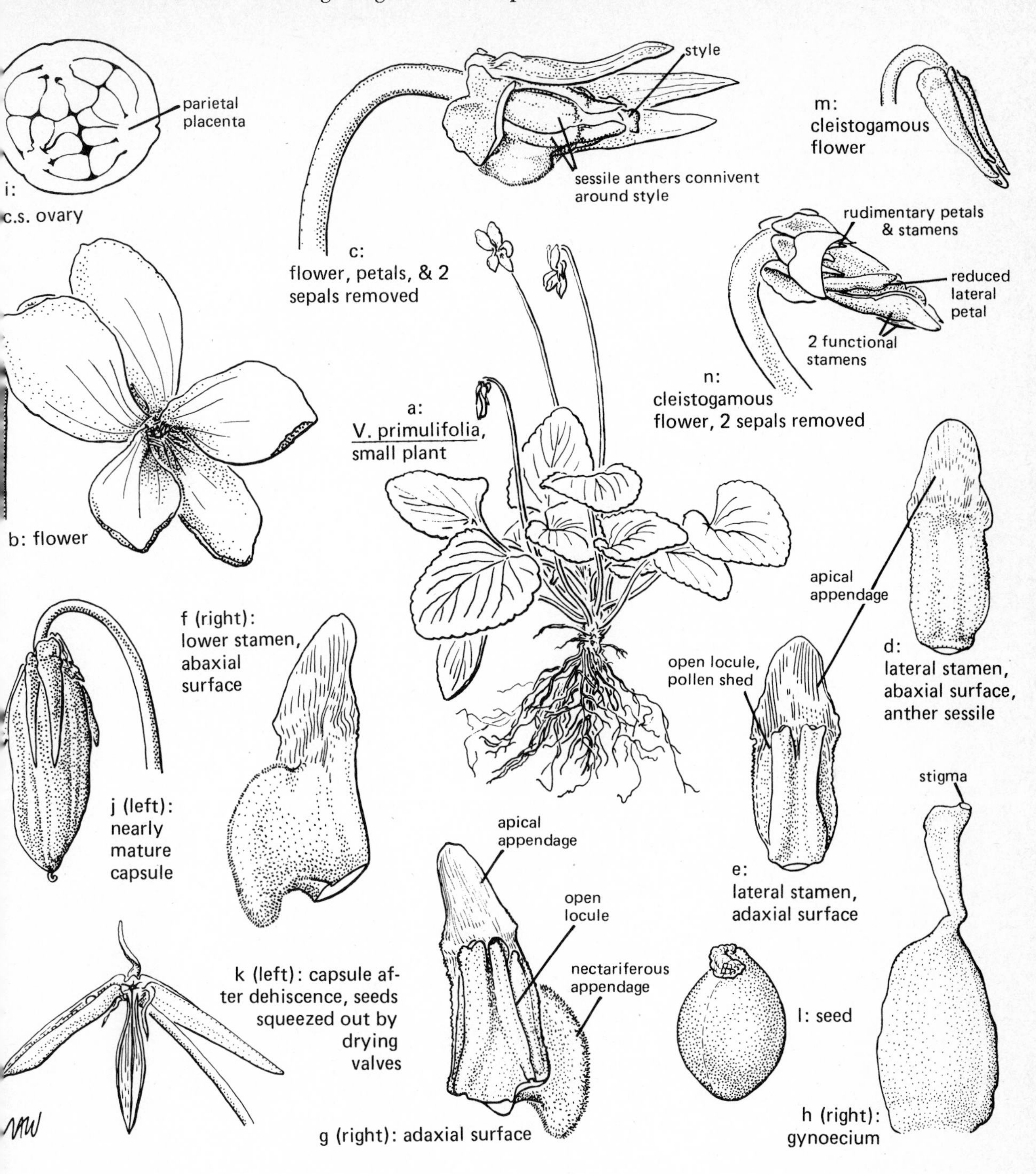

i:
c.s. ovary

parietal placenta

style

sessile anthers connivent around style

c:
flower, petals, & 2 sepals removed

m:
cleistogamous flower

rudimentary petals & stamens

reduced lateral petal

2 functional stamens

n:
cleistogamous flower, 2 sepals removed

b: flower

a:
V. primulifolia, small plant

apical appendage

d:
lateral stamen, abaxial surface, anther sessile

f (right):
lower stamen, abaxial surface

open locule, pollen shed

e:
lateral stamen, adaxial surface

j (left):
nearly mature capsule

stigma

apical appendage

open locule

nectariferous appendage

l: seed

k (left): capsule after dehiscence, seeds squeezed out by drying valves

g (right): adaxial surface

h (right):
gynoecium

71

PASSIFLORACEAE: Passiflora. a-i, P. incarnata

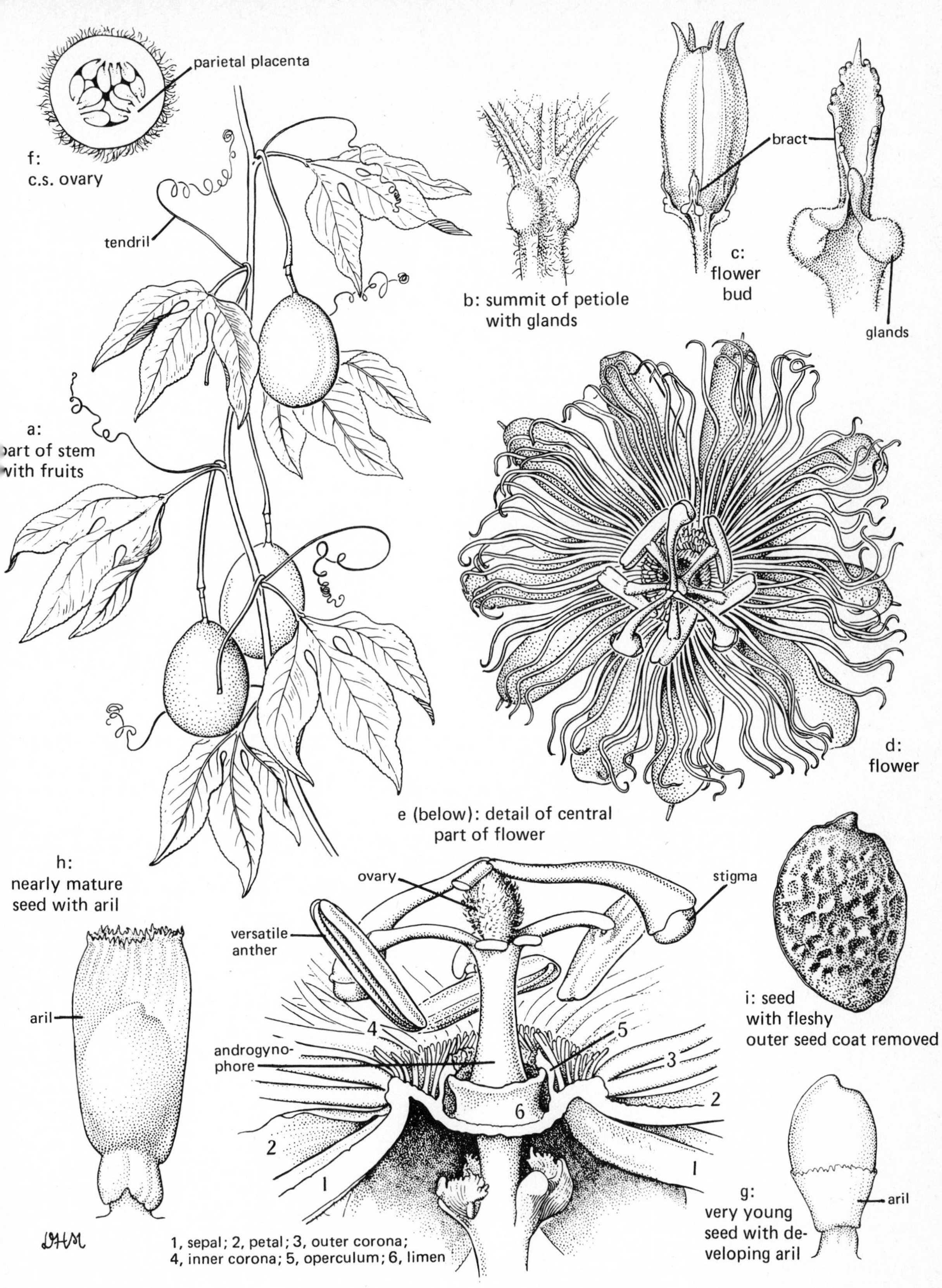

parietal placenta

f:
c.s. ovary

tendril

b: summit of petiole
with glands

bract

c:
flower
bud

glands

a:
part of stem
with fruits

d:
flower

e (below): detail of central
part of flower

h:
nearly mature
seed with aril

ovary

stigma

versatile
anther

i: seed
with fleshy
outer seed coat removed

aril

androgyno-
phore

g:
very young
seed with de-
veloping aril

aril

1, sepal; 2, petal; 3, outer corona;
4, inner corona; 5, operculum; 6, limen

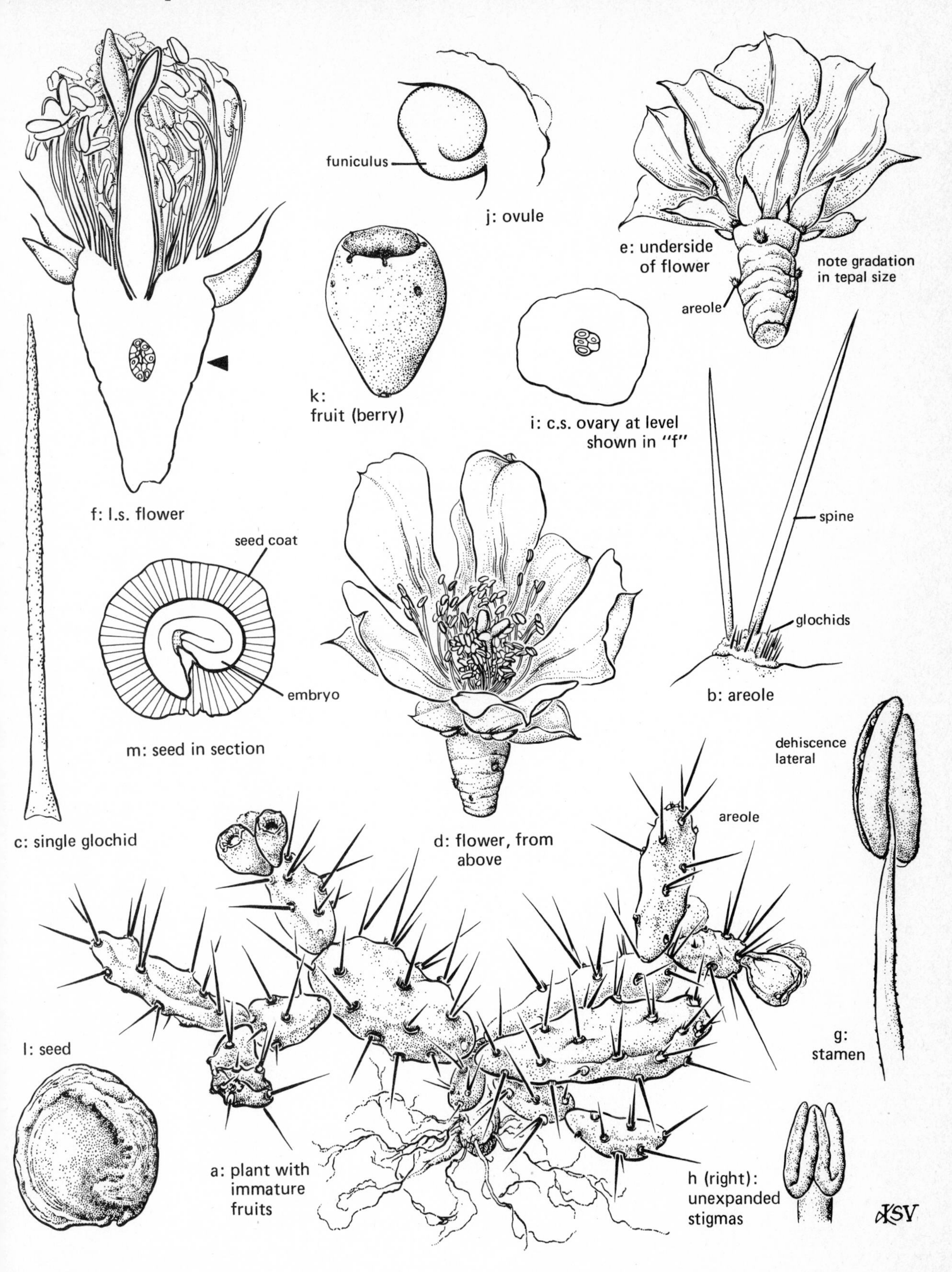

funiculus

j: ovule

e: underside
of flower

note gradation
in tepal size

areole

k:
fruit (berry)

i: c.s. ovary at level
shown in "f"

f: l.s. flower

spine

glochids

b: areole

seed coat

embryo

m: seed in section

dehiscence
lateral

c: single glochid

d: flower, from
above

areole

g:
stamen

l: seed

a: plant with
immature
fruits

h (right):
unexpanded
stigmas

KSV

73

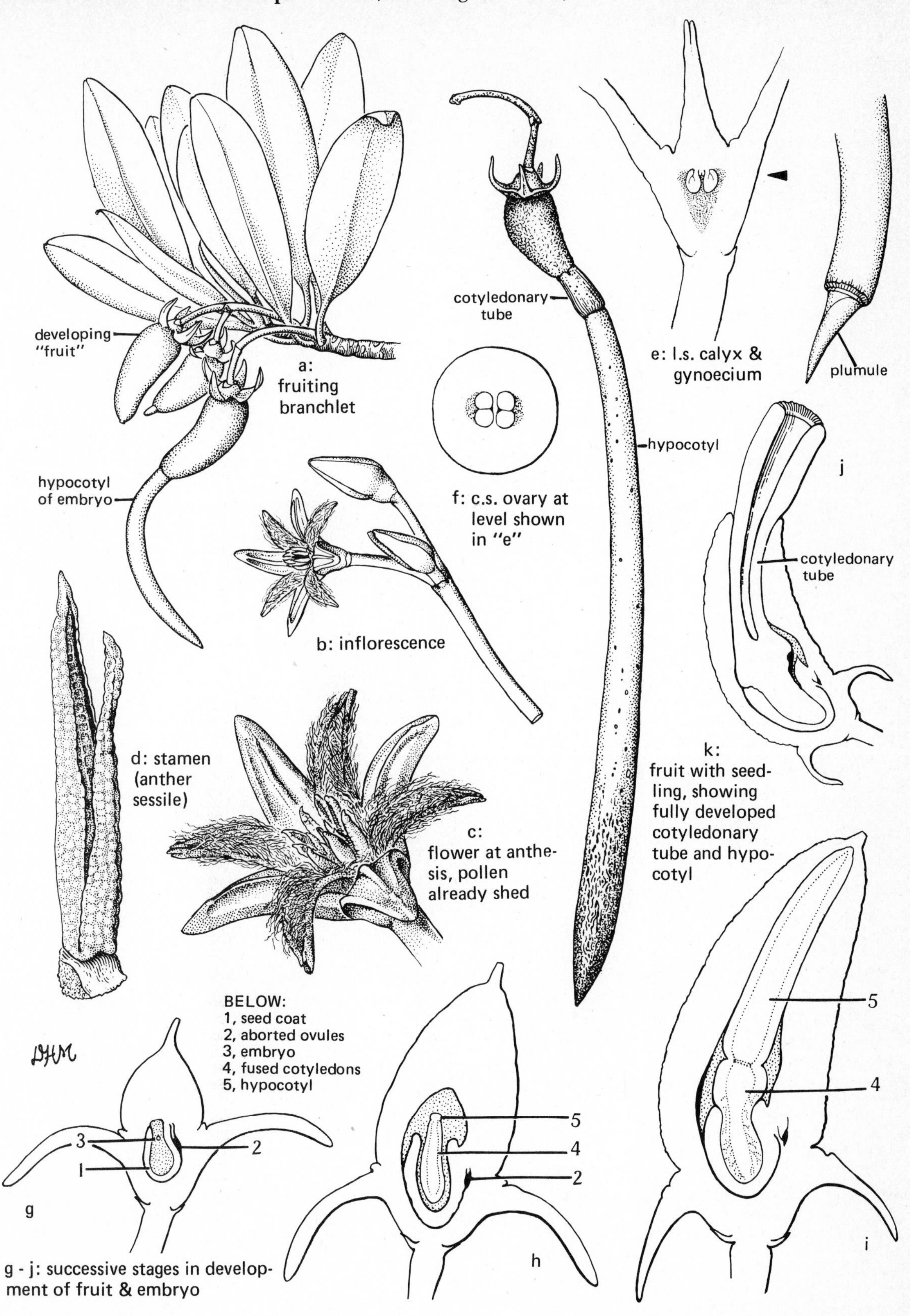

developing "fruit"

a: fruiting branchlet

hypocotyl of embryo

cotyledonary tube

f: c.s. ovary at level shown in "e"

b: inflorescence

d: stamen (anther sessile)

c: flower at anthesis, pollen already shed

e: l.s. calyx & gynoecium

plumule

hypocotyl

j

cotyledonary tube

k: fruit with seedling, showing fully developed cotyledonary tube and hypocotyl

BELOW:
1, seed coat
2, aborted ovules
3, embryo
4, fused cotyledons
5, hypocotyl

g

g - j: successive stages in development of fruit & embryo

h

i

5

4

3

2

1

5

4

2

5

4

DHM

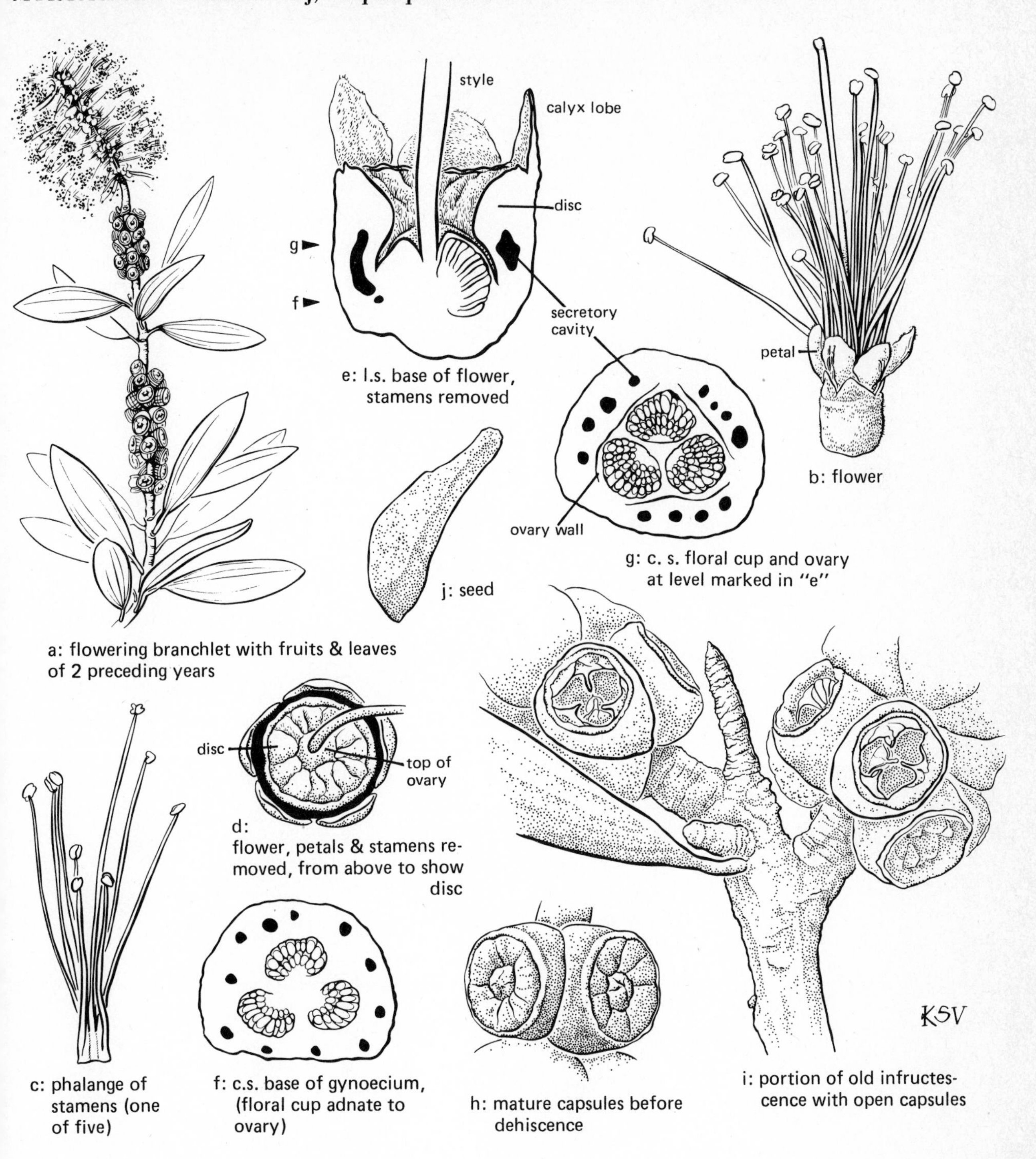

style

calyx lobe

disc

secretory cavity

g ►

f ►

e: l.s. base of flower, stamens removed

petal

b: flower

j: seed

ovary wall

g: c. s. floral cup and ovary at level marked in "e"

a: flowering branchlet with fruits & leaves of 2 preceding years

disc

top of ovary

d: flower, petals & stamens removed, from above to show disc

KSV

c: phalange of stamens (one of five)

f: c.s. base of gynoecium, (floral cup adnate to ovary)

h: mature capsules before dehiscence

i: portion of old infructescence with open capsules

MELASTOMATACEAE: Rhexia. a-j, R. virginica; k, R. nashii; l, R. nuttallii; m, R. petiolata; n, R. nuttallii; o, R. cubensis; p, R. aliphanus

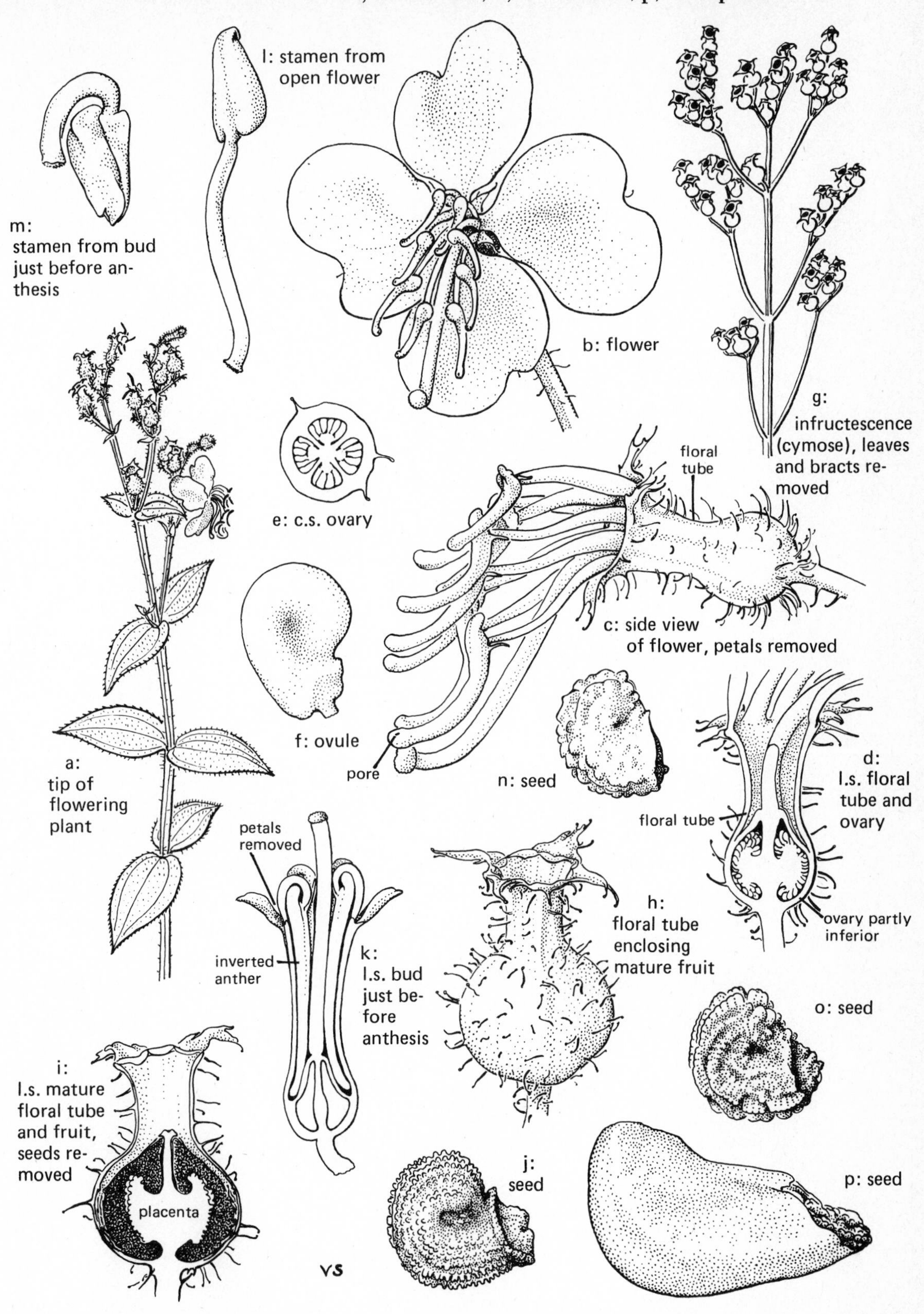

l: stamen from open flower

m: stamen from bud just before anthesis

b: flower

g: infructescence (cymose), leaves and bracts removed

e: c.s. ovary

floral tube

c: side view of flower, petals removed

f: ovule

pore

a: tip of flowering plant

n: seed

floral tube

d: l.s. floral tube and ovary

ovary partly inferior

petals removed

inverted anther

k: l.s. bud just before anthesis

h: floral tube enclosing mature fruit

o: seed

i: l.s. mature floral tube and fruit, seeds removed

placenta

j: seed

p: seed

vs

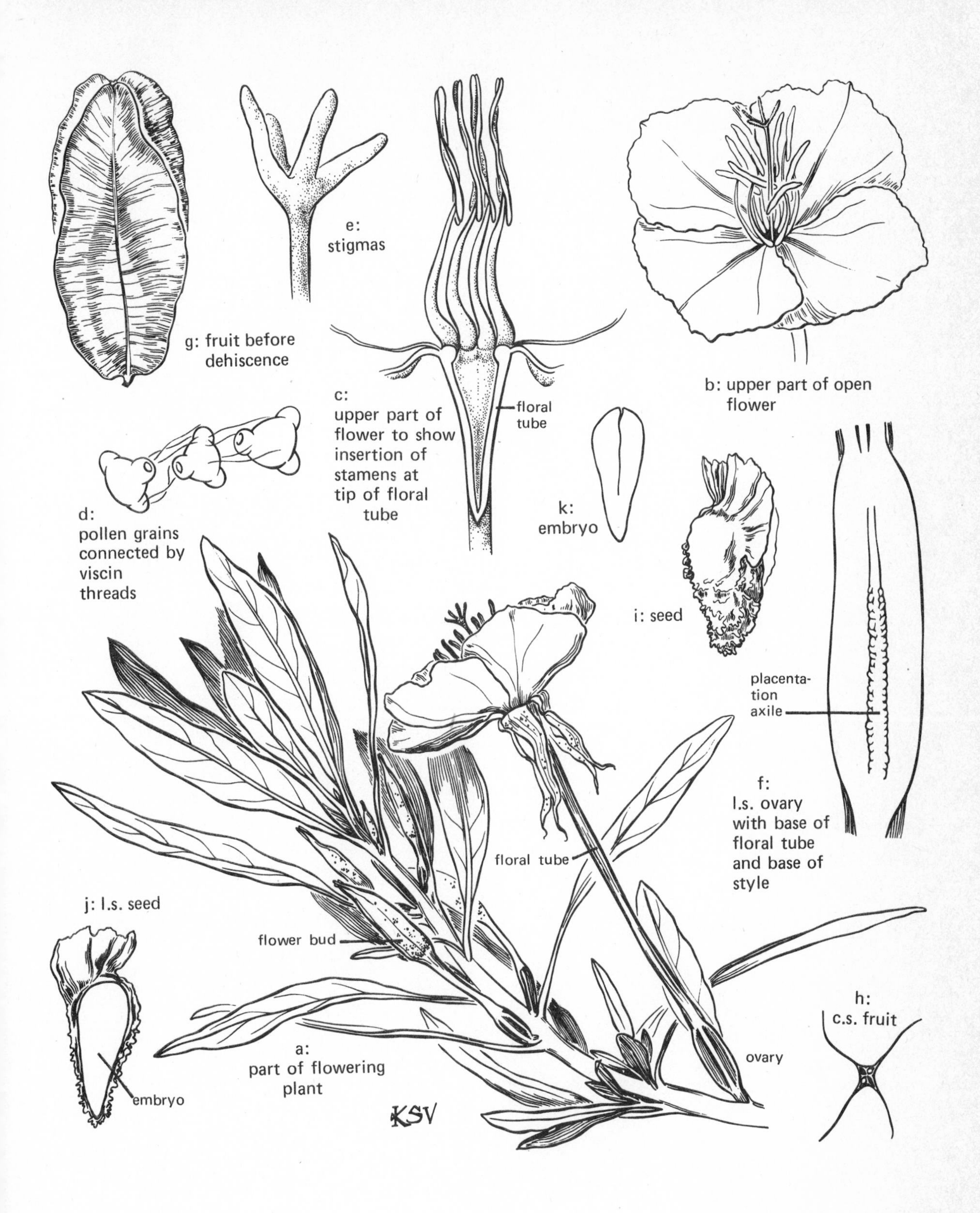

e: stigmas

g: fruit before dehiscence

c: upper part of flower to show insertion of stamens at tip of floral tube

floral tube

b: upper part of open flower

k: embryo

d: pollen grains connected by viscin threads

i: seed

placentation axile

f: l.s. ovary with base of floral tube and base of style

j: l.s. seed

floral tube

flower bud

embryo

a: part of flowering plant

ovary

h: c.s. fruit

KSV

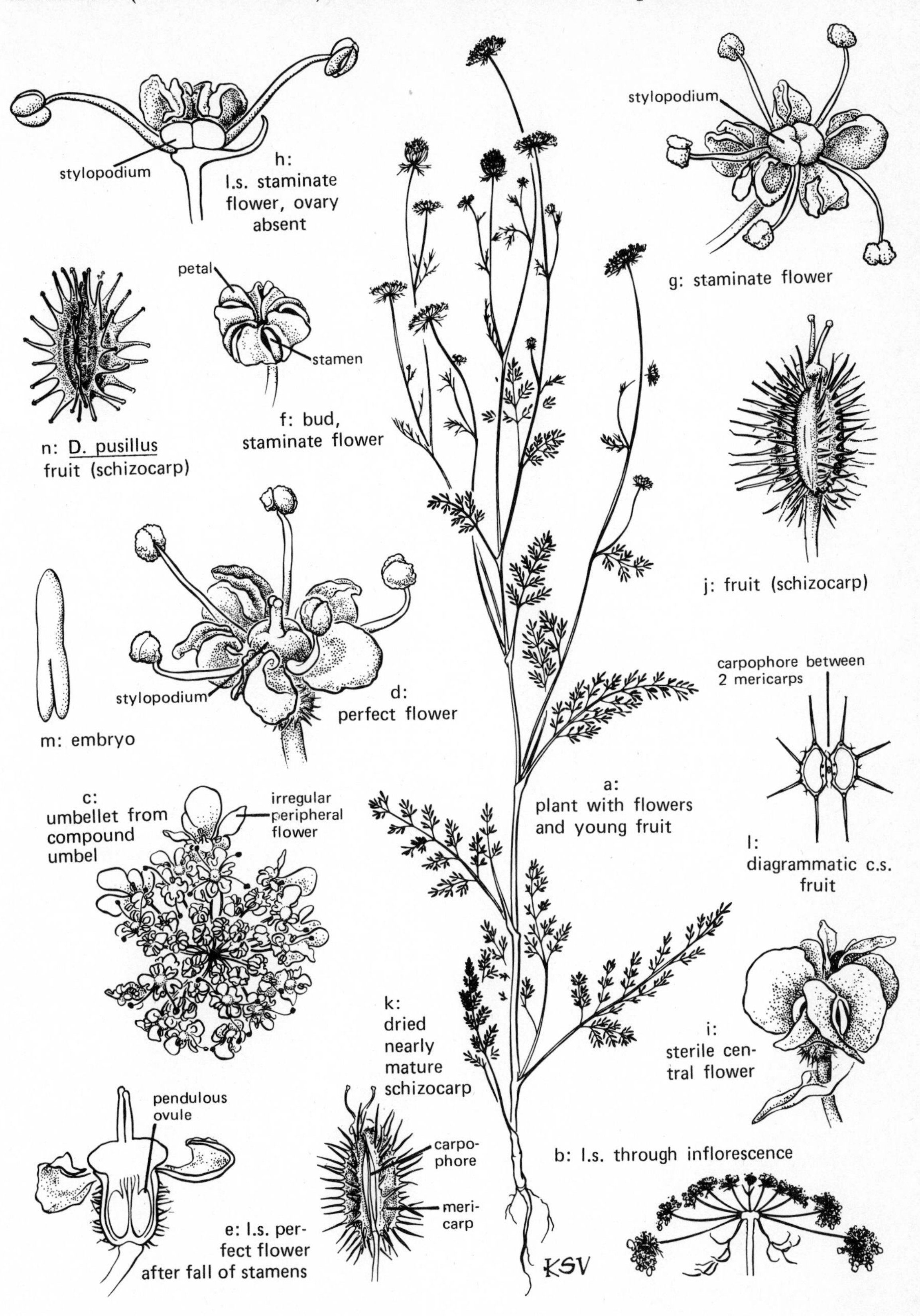

stylopodium

h:
l.s. staminate
flower, ovary
absent

petal

stamen

f: bud,
staminate flower

n: <u>D. pusillus</u>
fruit (schizocarp)

stylopodium

g: staminate flower

j: fruit (schizocarp)

m: embryo

stylopodium

d:
perfect flower

carpophore between
2 mericarps

l:
diagrammatic c.s.
fruit

c:
umbellet from
compound
umbel

irregular
peripheral
flower

a:
plant with flowers
and young fruit

i:
sterile cen-
tral flower

k:
dried
nearly
mature
schizocarp

pendulous
ovule

carpo-
phore

meri-
carp

e: l.s. per-
fect flower
after fall of stamens

b: l.s. through inflorescence

KSV

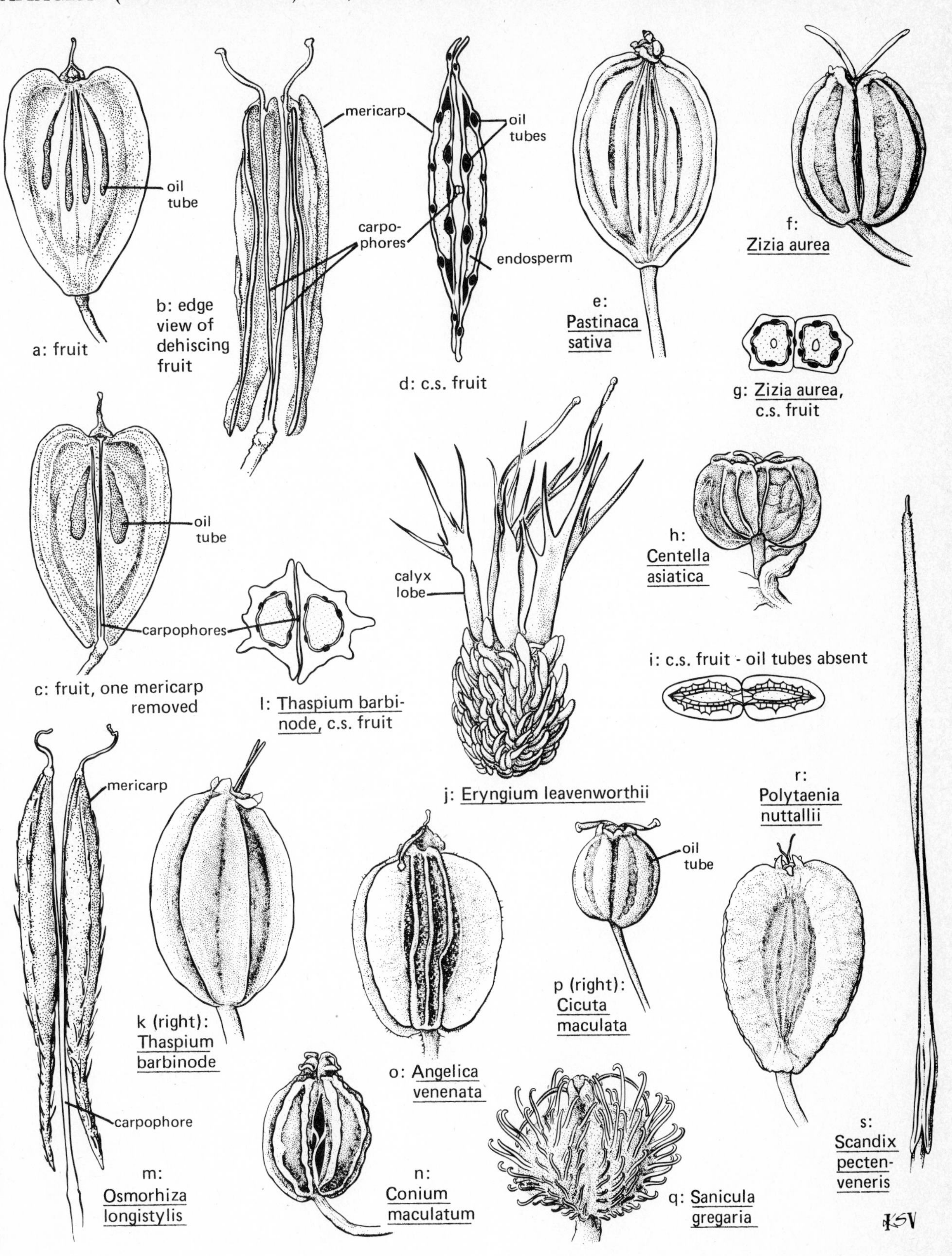

a: fruit

oil tube

b: edge view of dehiscing fruit

mericarp

carpophores

d: c.s. fruit

oil tubes

endosperm

e: Pastinaca sativa

f: Zizia aurea

g: Zizia aurea, c.s. fruit

oil tube

c: fruit, one mericarp removed

carpophores

l: Thaspium barbinode, c.s. fruit

calyx lobe

j: Eryngium leavenworthii

h: Centella asiatica

i: c.s. fruit - oil tubes absent

r: Polytaenia nuttallii

mericarp

carpophore

m: Osmorhiza longistylis

k (right): Thaspium barbinode

n: Conium maculatum

o: Angelica venenata

p (right): Cicuta maculata

oil tube

q: Sanicula gregaria

s: Scandix pecten- veneris

KSV

79

CORNACEAE: Cornus. a-e, C. amomum; f-h, C. florida; i, C. alternifolia

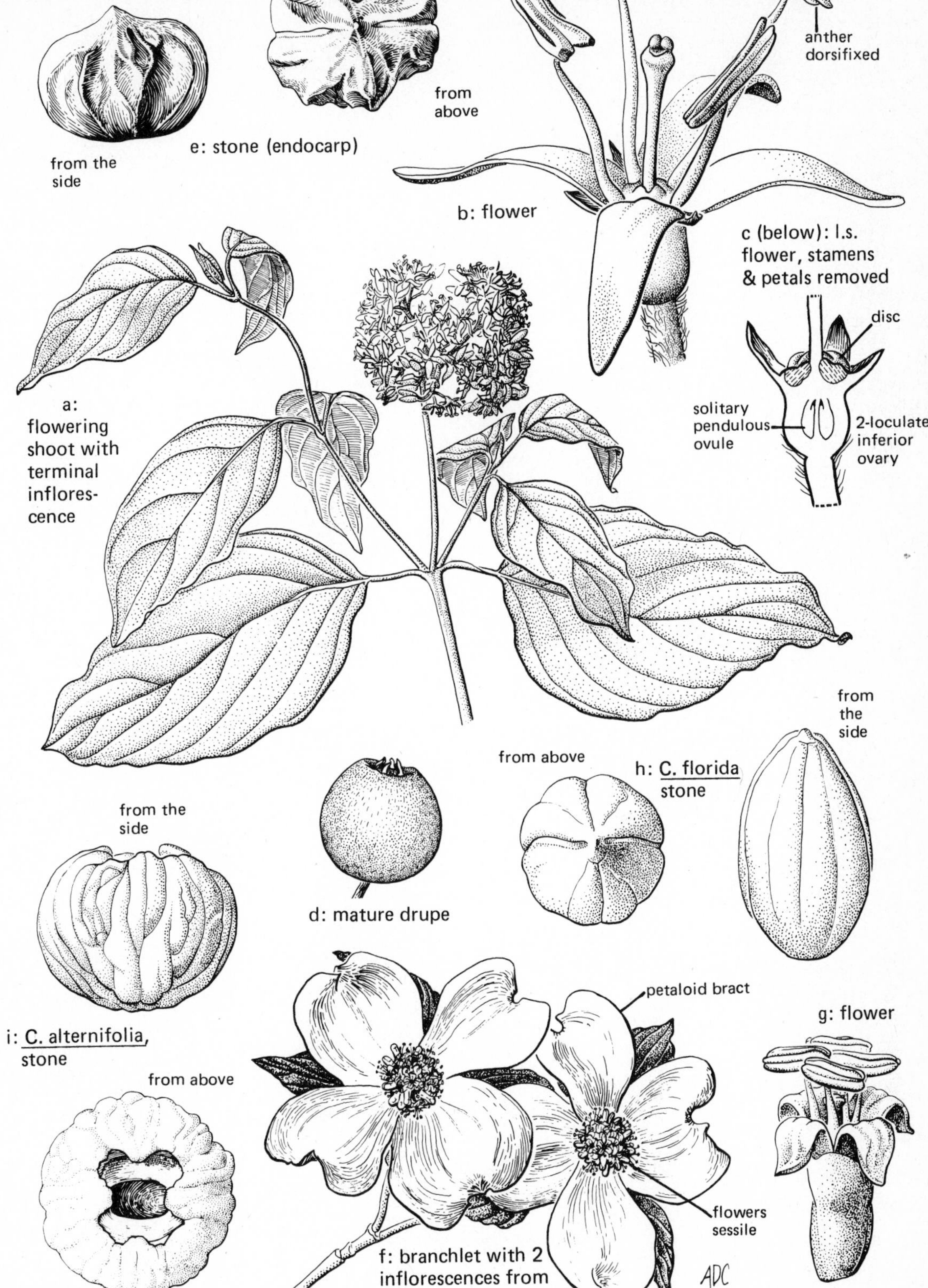

from the side

e: stone (endocarp)

from above

b: flower

anther dorsifixed

c (below): l.s. flower, stamens & petals removed

disc

a: flowering shoot with terminal inflorescence

solitary pendulous ovule

2-loculate inferior ovary

from the side

from above

h: C. florida stone

from the side

d: mature drupe

i: C. alternifolia, stone

petaloid bract

g: flower

from above

flowers sessile

f: branchlet with 2 inflorescences from above

ADC

80

ERICACEAE: Rhododendron. a-e, R. carolinianum; f-k, R. vaseyi; l-q, R. atlanticum

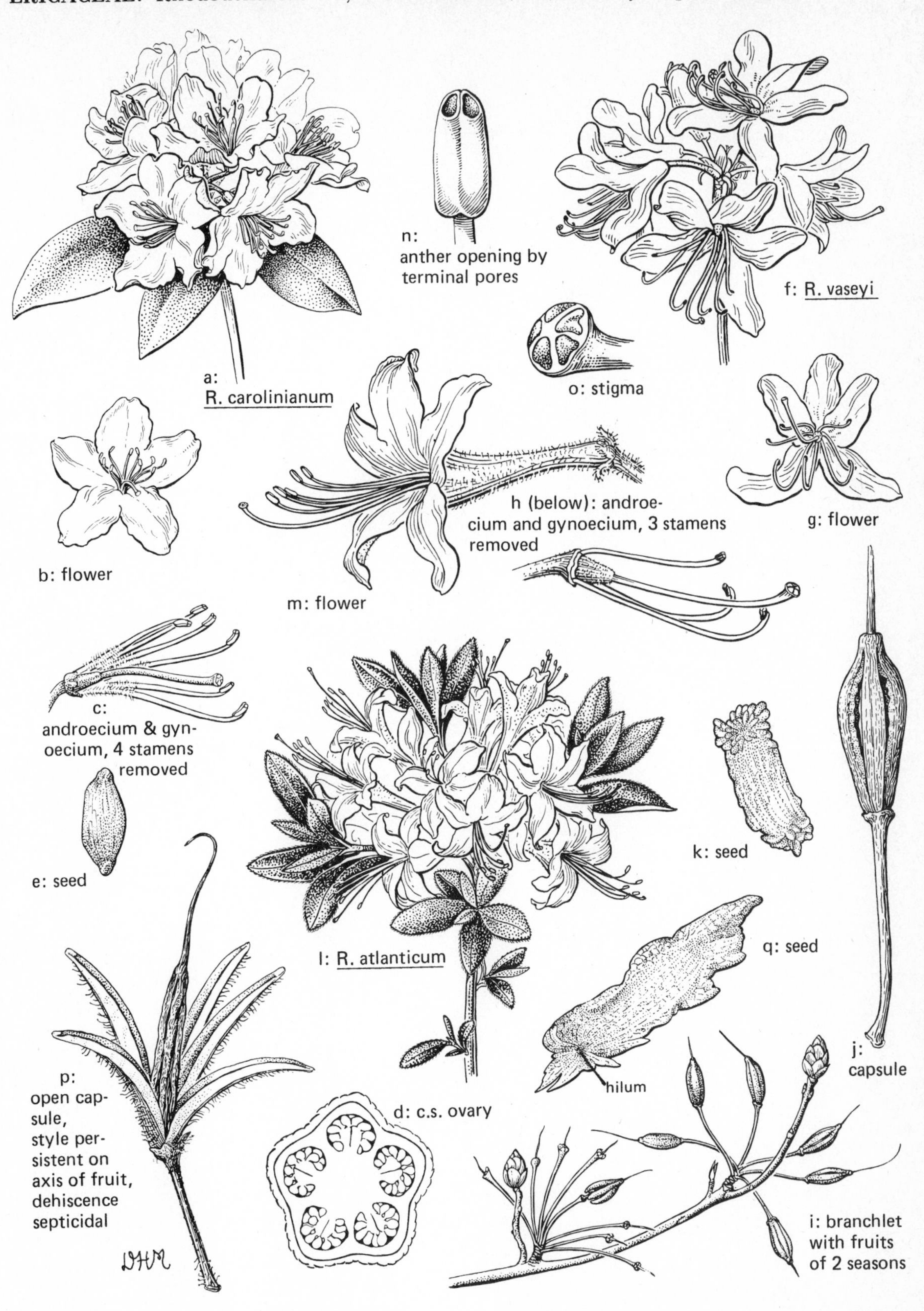

n:
anther opening by
terminal pores

a:
R. carolinianum

o: stigma

f: R. vaseyi

h (below): androe-
cium and gynoecium, 3 stamens
removed

g: flower

b: flower

m: flower

c:
androecium & gyn-
oecium, 4 stamens
removed

k: seed

e: seed

l: R. atlanticum

q: seed

j:
capsule

p:
open cap-
sule,
style per-
sistent on
axis of fruit,
dehiscence
septicidal

hilum

d: c.s. ovary

i: branchlet
with fruits
of 2 seasons

DHM

ERICACEAE: Oxydendrum. a-g, O. arboreum

a: flowering branch

b: flower

c: l.s. flower

d: three views of stamens

anther opening introrsely by an elongated pore

e: detail of raceme with erect immature fruits

f: open capsule, 1 valve removed dehiscence loculicidal

g: seed

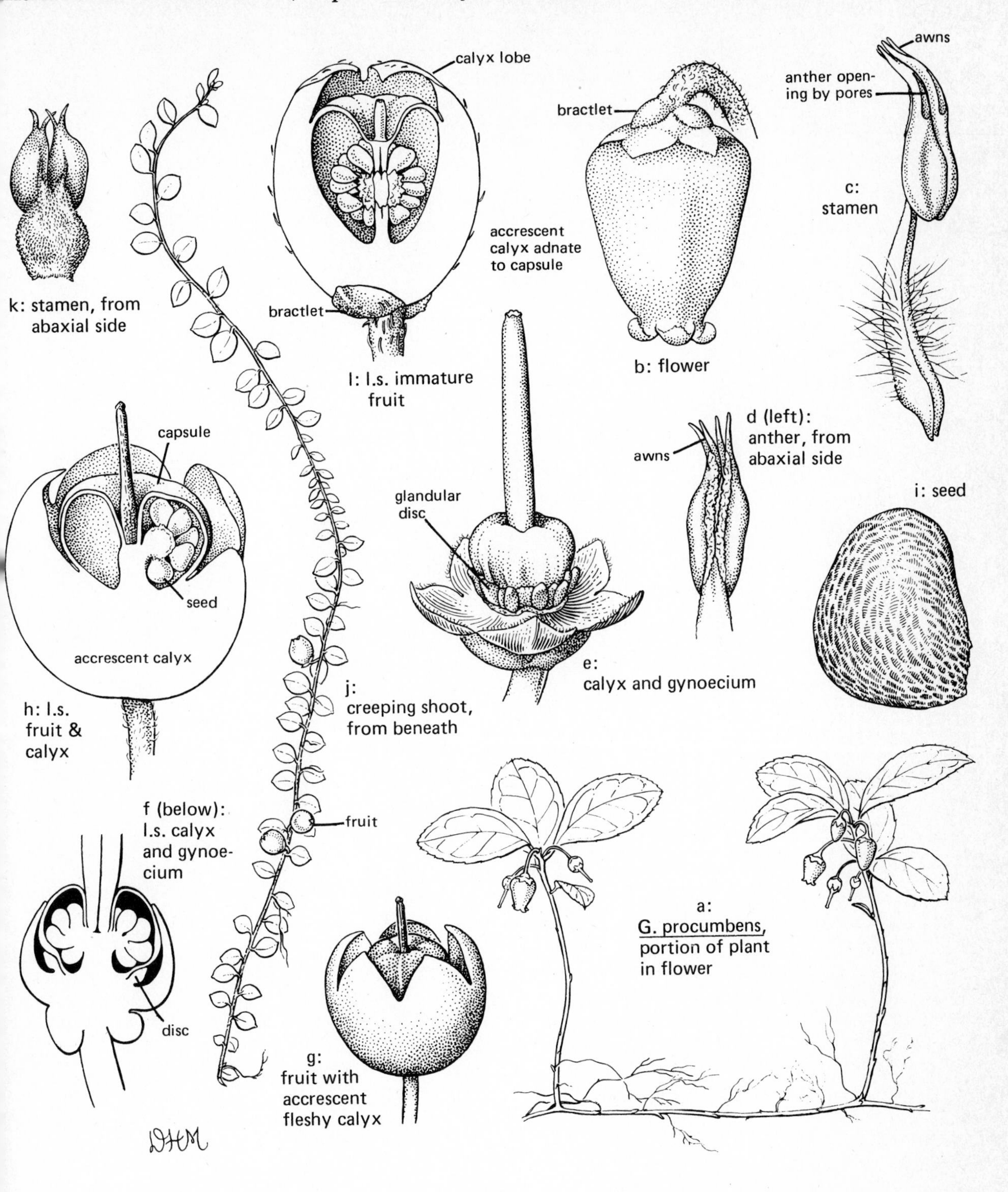

k: stamen, from
abaxial side

calyx lobe

accrescent
calyx adnate
to capsule

bractlet

l: l.s. immature
fruit

bractlet

b: flower

awns

anther open-
ing by pores

c:
stamen

capsule

seed

accrescent calyx

h: l.s.
fruit &
calyx

glandular
disc

j:
creeping shoot,
from beneath

d (left):
anther, from
abaxial side

awns

e:
calyx and gynoecium

i: seed

f (below):
l.s. calyx
and gynoe-
cium

fruit

disc

g:
fruit with
accrescent
fleshy calyx

a:
G. procumbens,
portion of plant
in flower

DHM

ERICACEAE: Chimaphila. a-i, C. maculata; j, C. umbellata var. cisatlantica

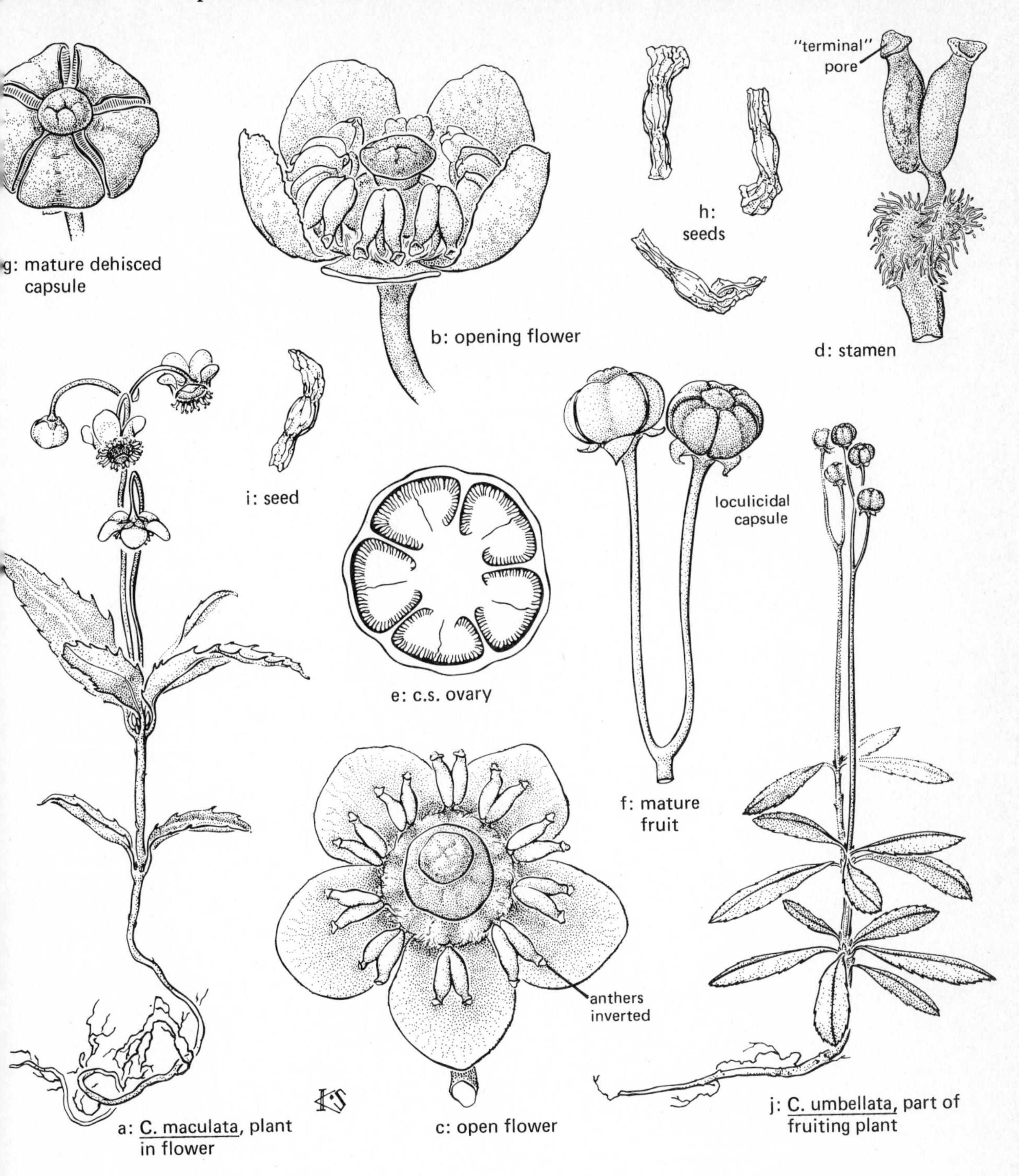

g: mature dehisced capsule

b: opening flower

h: seeds

d: stamen

i: seed

e: c.s. ovary

loculicidal capsule

"terminal" pore

f: mature fruit

anthers inverted

a: C. maculata, plant in flower

c: open flower

j: C. umbellata, part of fruiting plant

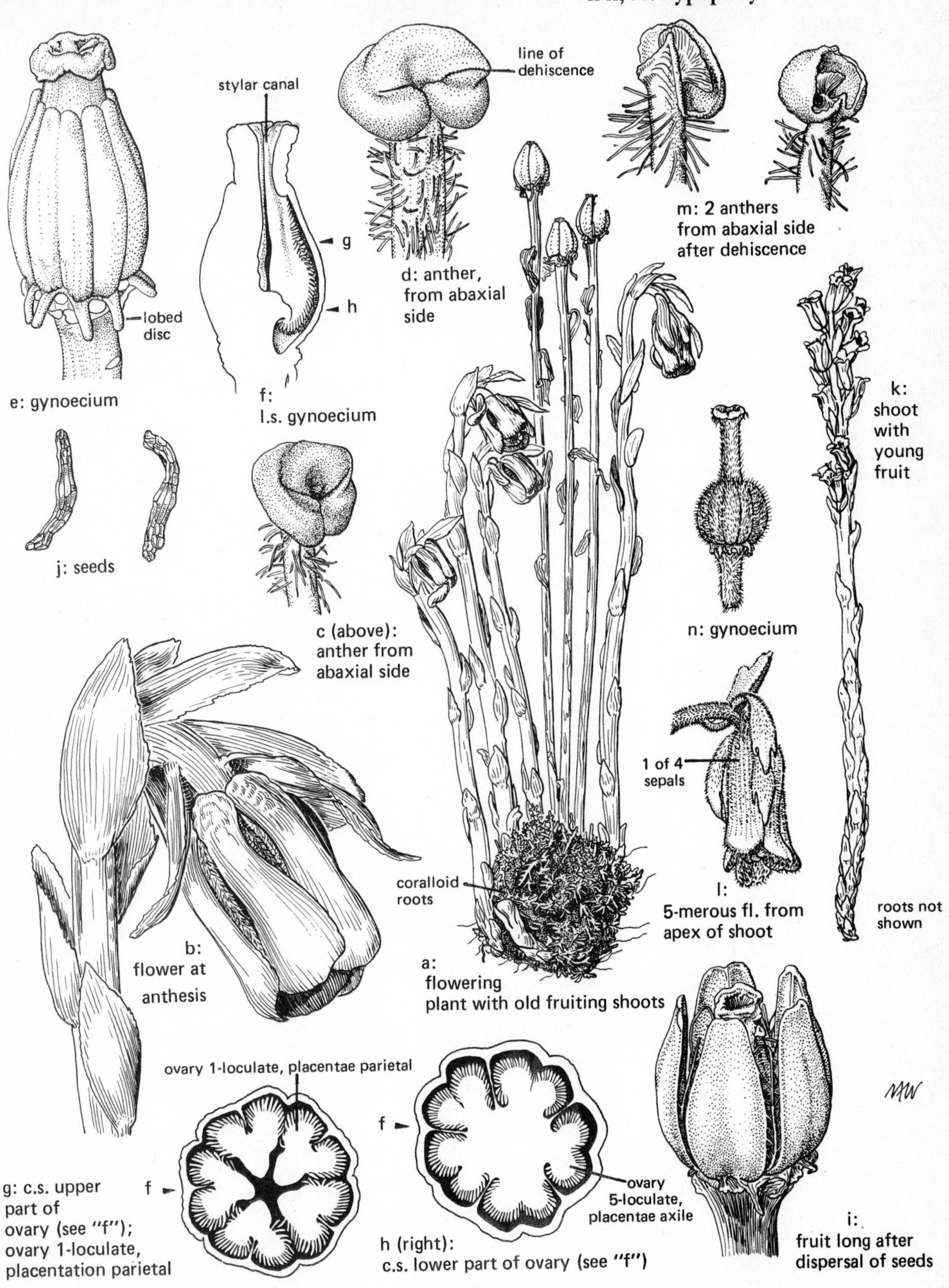

line of dehiscence

m: 2 anthers from abaxial side after dehiscence

d: anther, from abaxial side

stylar canal

g

h

f: l.s. gynoecium

e: gynoecium

lobed disc

j: seeds

c (above): anther from abaxial side

k: shoot with young fruit

n: gynoecium

1 of 4 sepals

l: 5-merous fl. from apex of shoot

roots not shown

b: flower at anthesis

coralloid roots

a: flowering plant with old fruiting shoots

ovary 1-loculate, placentae parietal

f

f

ovary 5-loculate, placentae axile

g: c.s. upper part of ovary (see "f''); ovary 1-loculate, placentation parietal

h (right): c.s. lower part of ovary (see "f'')

i: fruit long after dispersal of seeds

85

PRIMULACEAE: Dodecatheon. a–l, D. media subsp. media

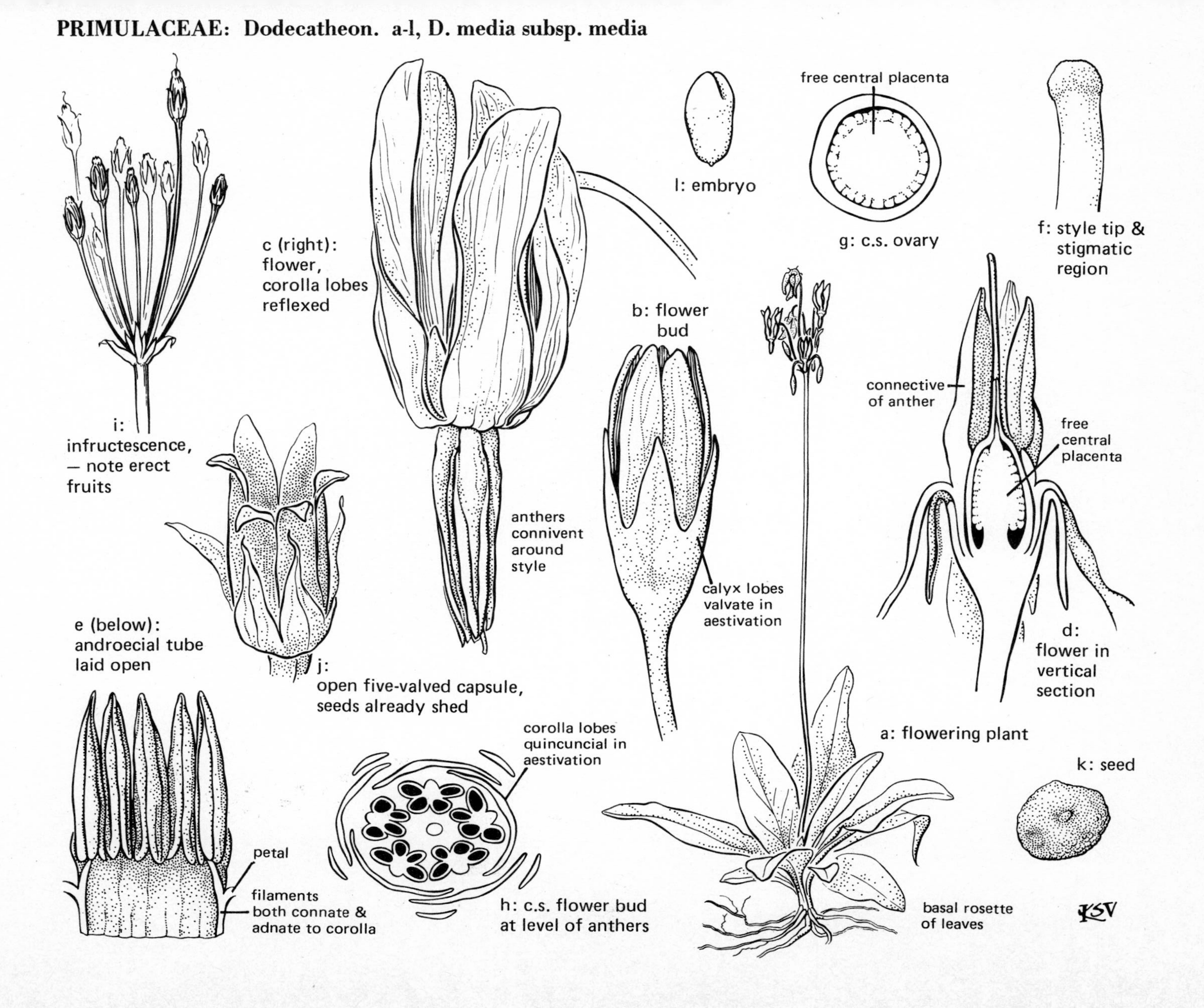

f: style tip & stigmatic region

free central placenta

connective of anther

d: flower in vertical section

k: seed

free central placenta

g: c.s. ovary

a: flowering plant

basal rosette of leaves

l: embryo

b: flower bud

calyx lobes valvate in aestivation

c (right): flower, corolla lobes reflexed

anthers connivent around style

corolla lobes quincuncial in aestivation

h: c.s. flower bud at level of anthers

j: open five-valved capsule, seeds already shed

i: infructescence, — note erect fruits

e (below): androecial tube laid open

petal

filaments both connate & adnate to corolla

OLEACEAE: Chionanthus. a–k, C. virginicus; l, C. pygmaeus

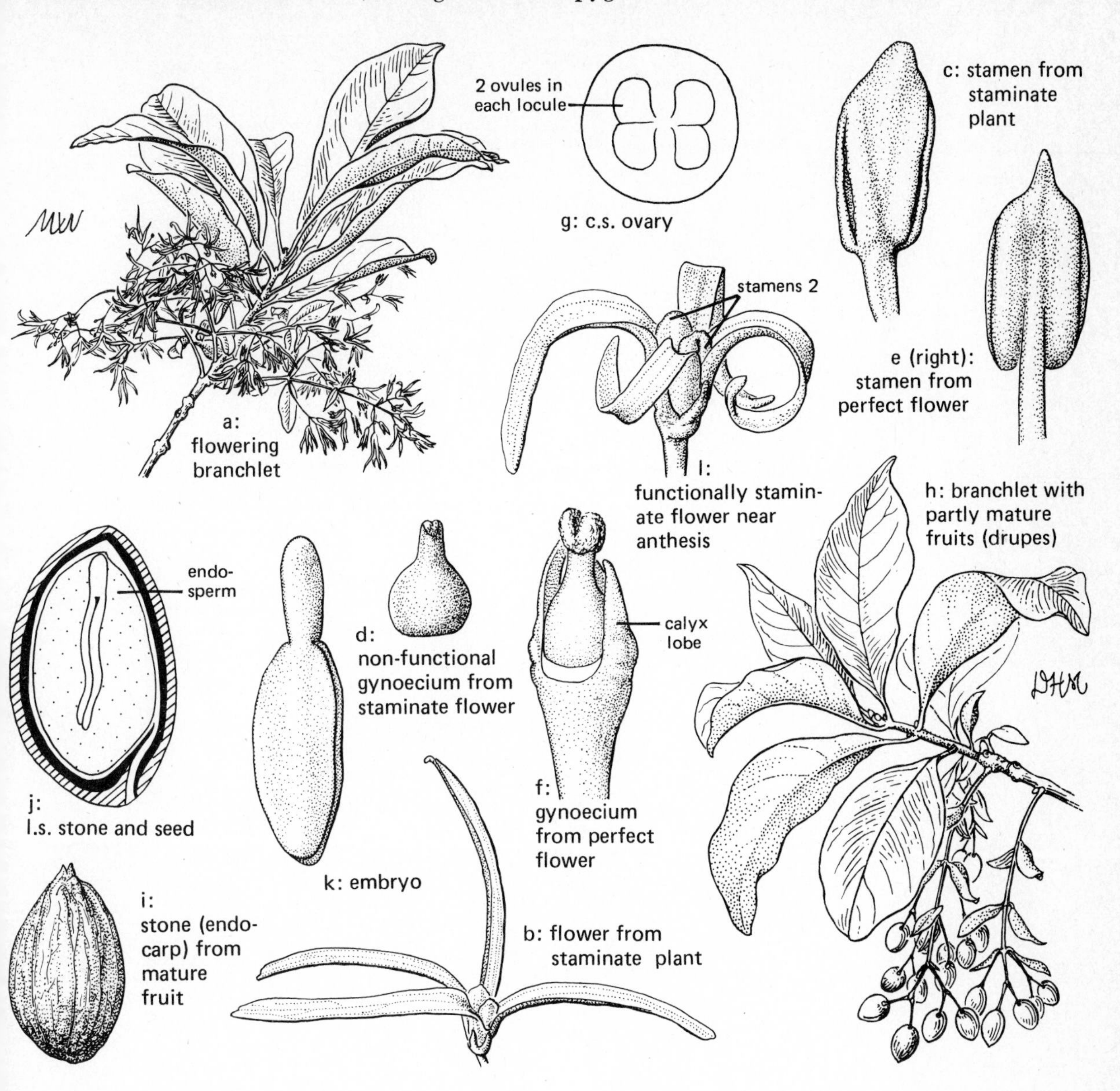

2 ovules in each locule

g: c.s. ovary

c: stamen from staminate plant

stamens 2

e (right): stamen from perfect flower

a: flowering branchlet

l: functionally staminate flower near anthesis

h: branchlet with partly mature fruits (drupes)

endo-sperm

d: non-functional gynoecium from staminate flower

calyx lobe

j: l.s. stone and seed

f: gynoecium from perfect flower

i: stone (endo-carp) from mature fruit

k: embryo

b: flower from staminate plant

GENTIANACEAE: Sabatia. a-g, S. kennedyana; h, S. campanulata; i, S. difformis; j, S. gentianoides

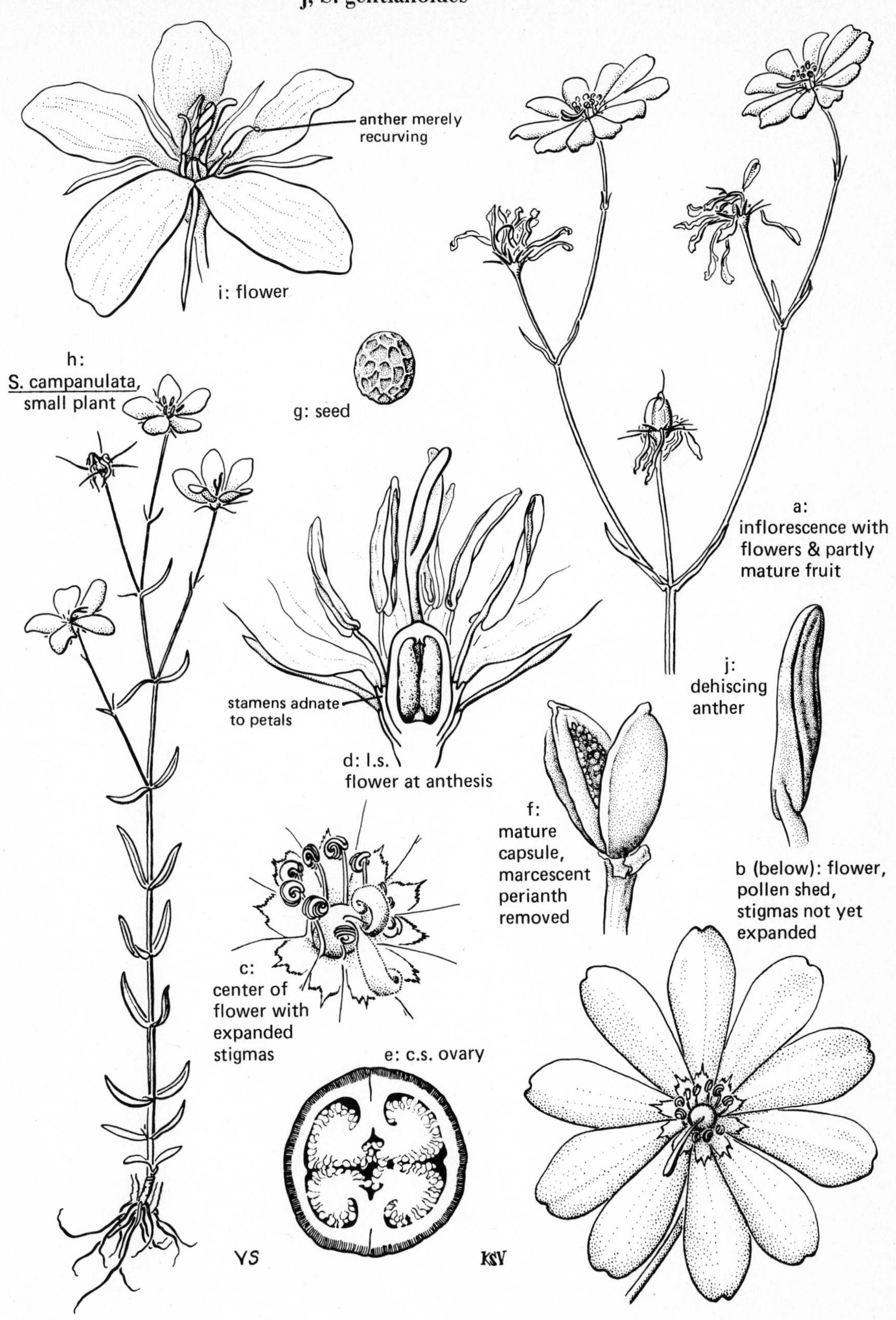

anther merely recurving

i: flower

h:
S. campanulata,
small plant

g: seed

a:
inflorescence with
flowers & partly
mature fruit

stamens adnate
to petals

d: l.s.
flower at anthesis

j:
dehiscing
anther

f:
mature
capsule,
marcescent
perianth
removed

b (below): flower,
pollen shed,
stigmas not yet
expanded

c:
center of
flower with
expanded
stigmas

e: c.s. ovary

YS

KSV

APOCYNACEAE: Amsonia. a-l, A. tabernaemontana

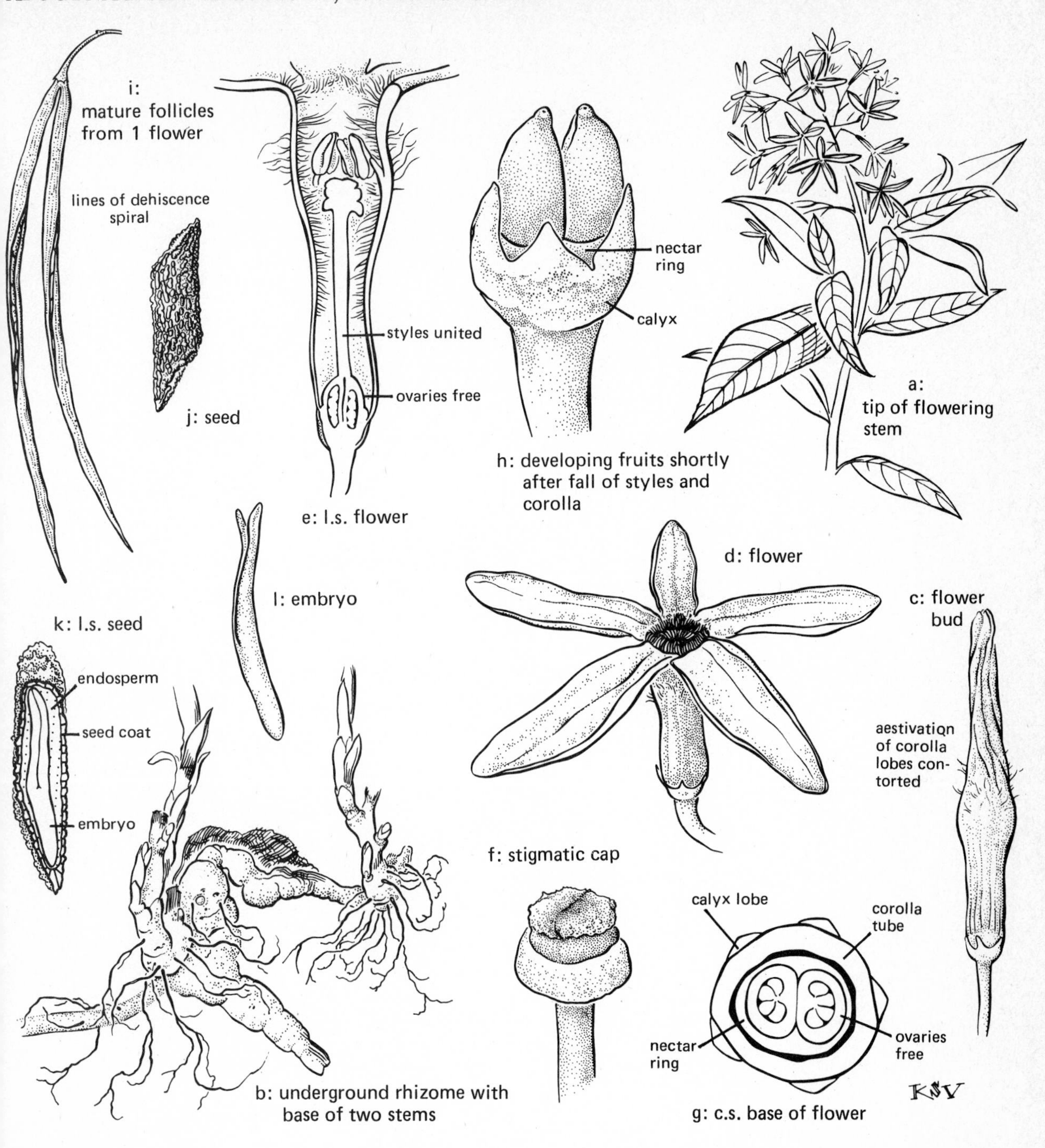

i: mature follicles from 1 flower

lines of dehiscence spiral

j: seed

styles united

ovaries free

e: l.s. flower

nectar ring

calyx

h: developing fruits shortly after fall of styles and corolla

a: tip of flowering stem

l: embryo

d: flower

c: flower bud

k: l.s. seed

endosperm

seed coat

embryo

aestivation of corolla lobes contorted

f: stigmatic cap

calyx lobe

corolla tube

nectar ring

ovaries free

b: underground rhizome with base of two stems

g: c.s. base of flower

KSV

ASCLEPIADACEAE: Asclepias. a-f, A. syriaca; g, A. connivens; h, A. pedicellata

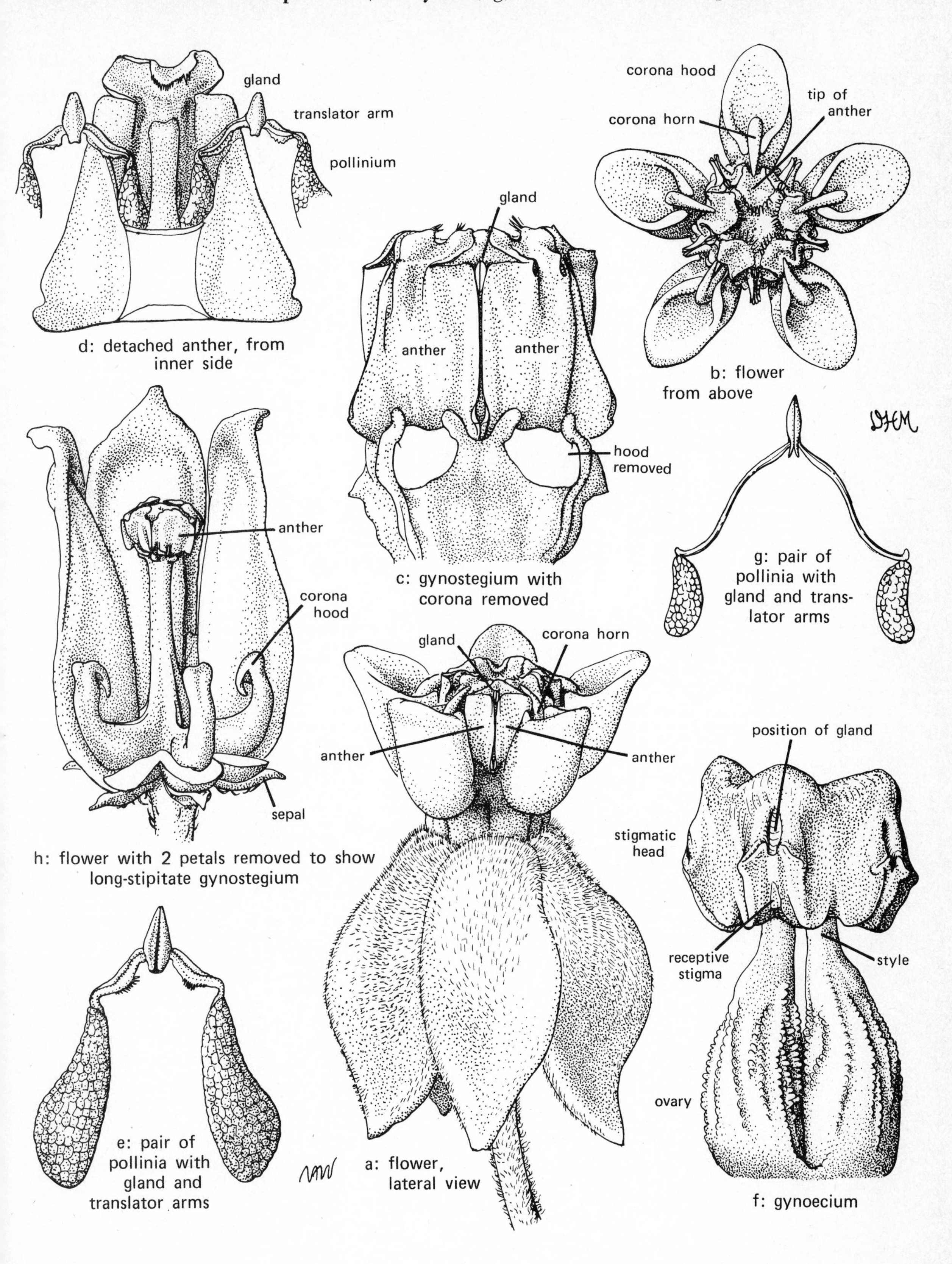

gland

translator arm

pollinium

d: detached anther, from
inner side

gland

anther   anther

hood
removed

c: gynostegium with
corona removed

corona hood

corona horn

tip of
anther

b: flower
from above

g: pair of
pollinia with
gland and trans-
lator arms

anther

corona
hood

sepal

h: flower with 2 petals removed to show
long-stipitate gynostegium

gland   corona horn

anther   anther

stigmatic
head

position of gland

receptive
stigma

style

ovary

e: pair of
pollinia with
gland and
translator arms

a: flower,
lateral view

f: gynoecium

ASCLEPIADACEAE: Asclepias. a, b, A. syriaca; c-f, A. incarnata

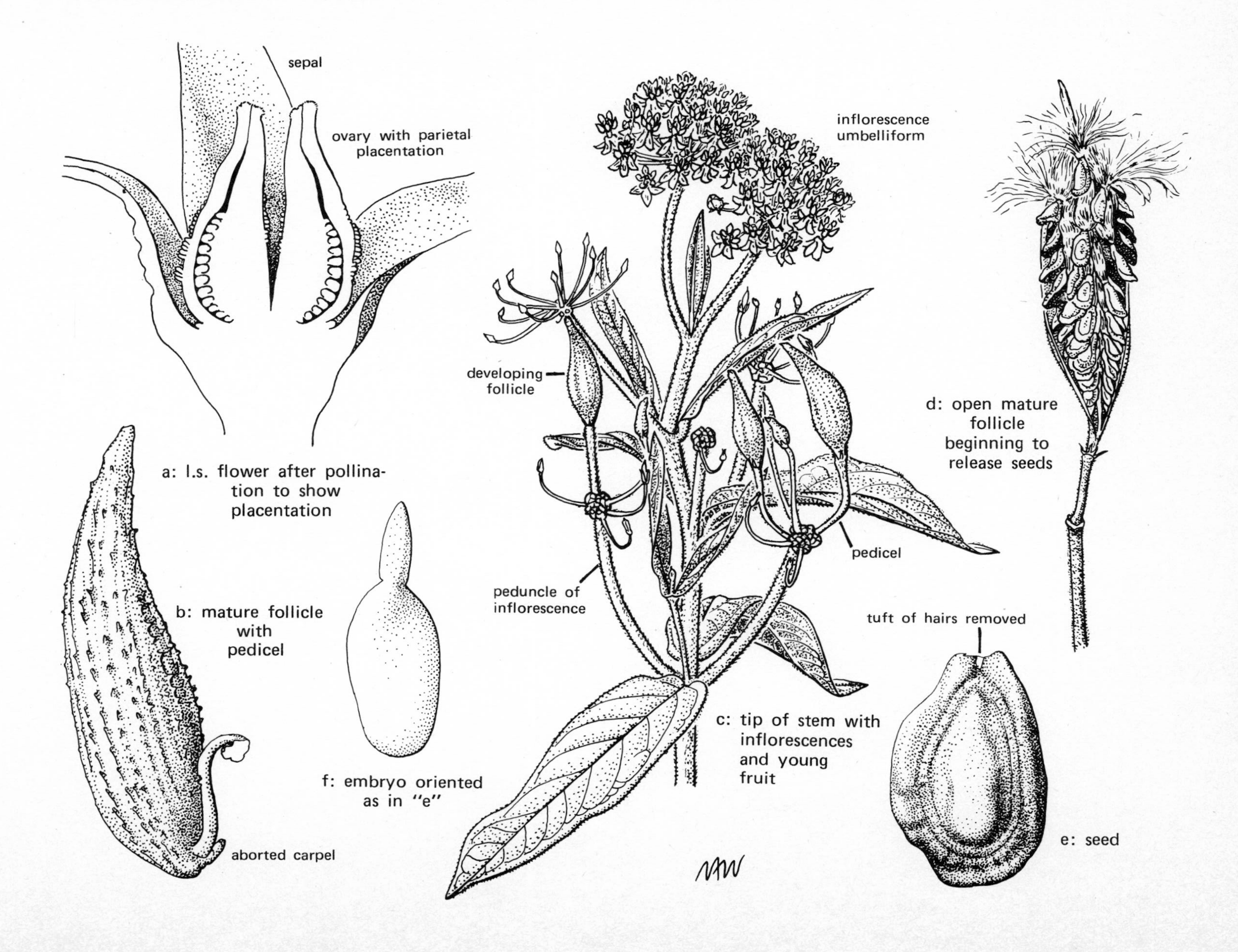

d: open mature follicle beginning to release seeds

inflorescence umbelliform

pedicel

tuft of hairs removed

e: seed

c: tip of stem with inflorescences and young fruit

developing follicle

peduncle of inflorescence

sepal

ovary with parietal placentation

a: l.s. flower after pollination to show placentation

b: mature follicle with pedicel

f: embryo oriented as in "e"

aborted carpel

CONVOLVULACEAE: Calystegia. a-l, C. sepium; m, C. spithamaea

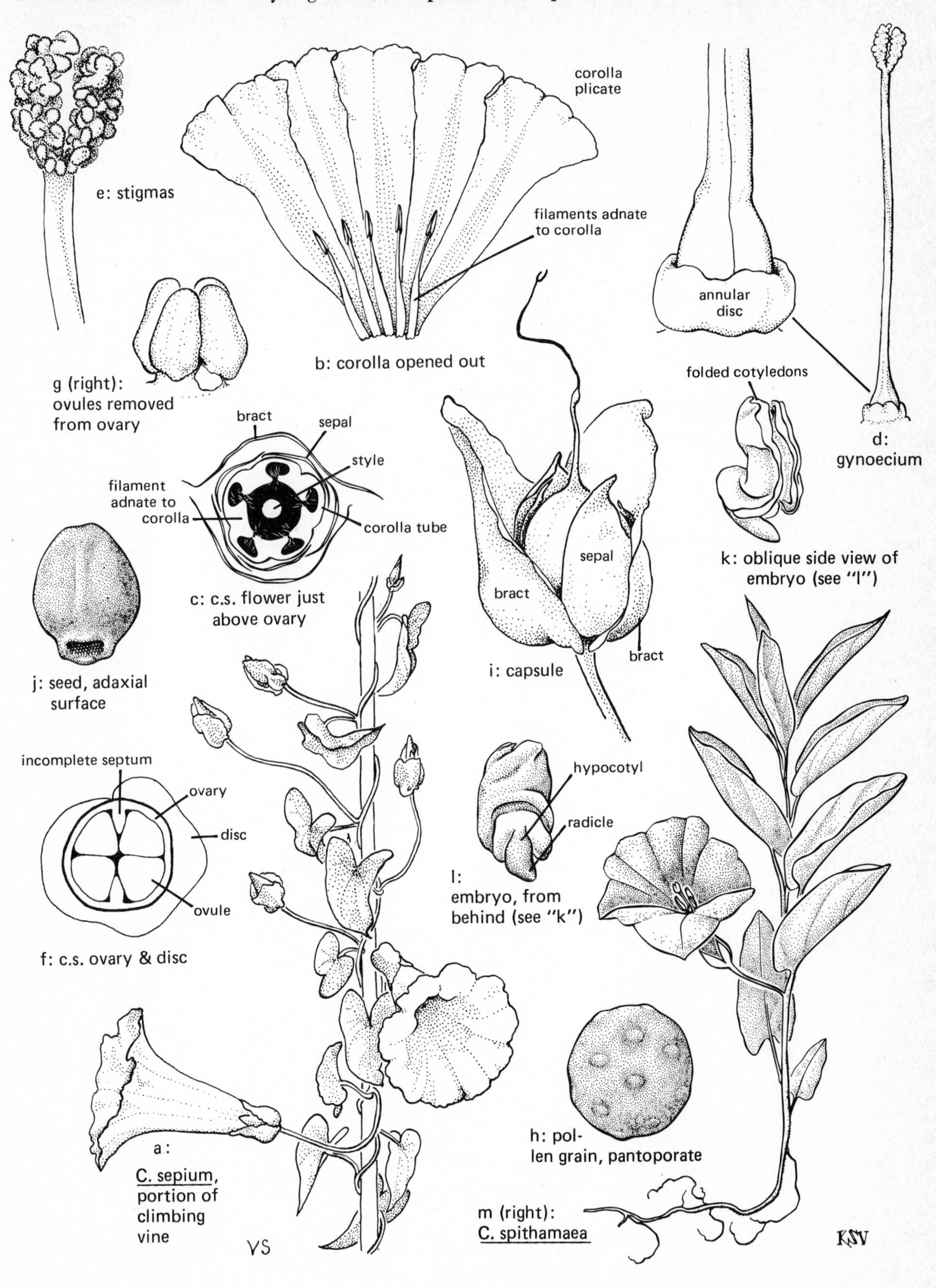

e: stigmas

corolla plicate

filaments adnate to corolla

b: corolla opened out

g (right): ovules removed from ovary

annular disc

folded cotyledons

d: gynoecium

bract
sepal
style

filament adnate to corolla

corolla tube

c: c.s. flower just above ovary

k: oblique side view of embryo (see "l")

sepal

bract

bract

i: capsule

j: seed, adaxial surface

incomplete septum

ovary

disc

ovule

f: c.s. ovary & disc

hypocotyl

radicle

l: embryo, from behind (see "k")

h: pollen grain, pantoporate

a: C. sepium, portion of climbing vine

VS

m (right): C. spithamaea

KSV

BORAGINACEAE: **Heliotropium.** **a-f, H. curassavicum; g-k, H. leavenworthii; l, m, H. indicum; n, H. tenellum**

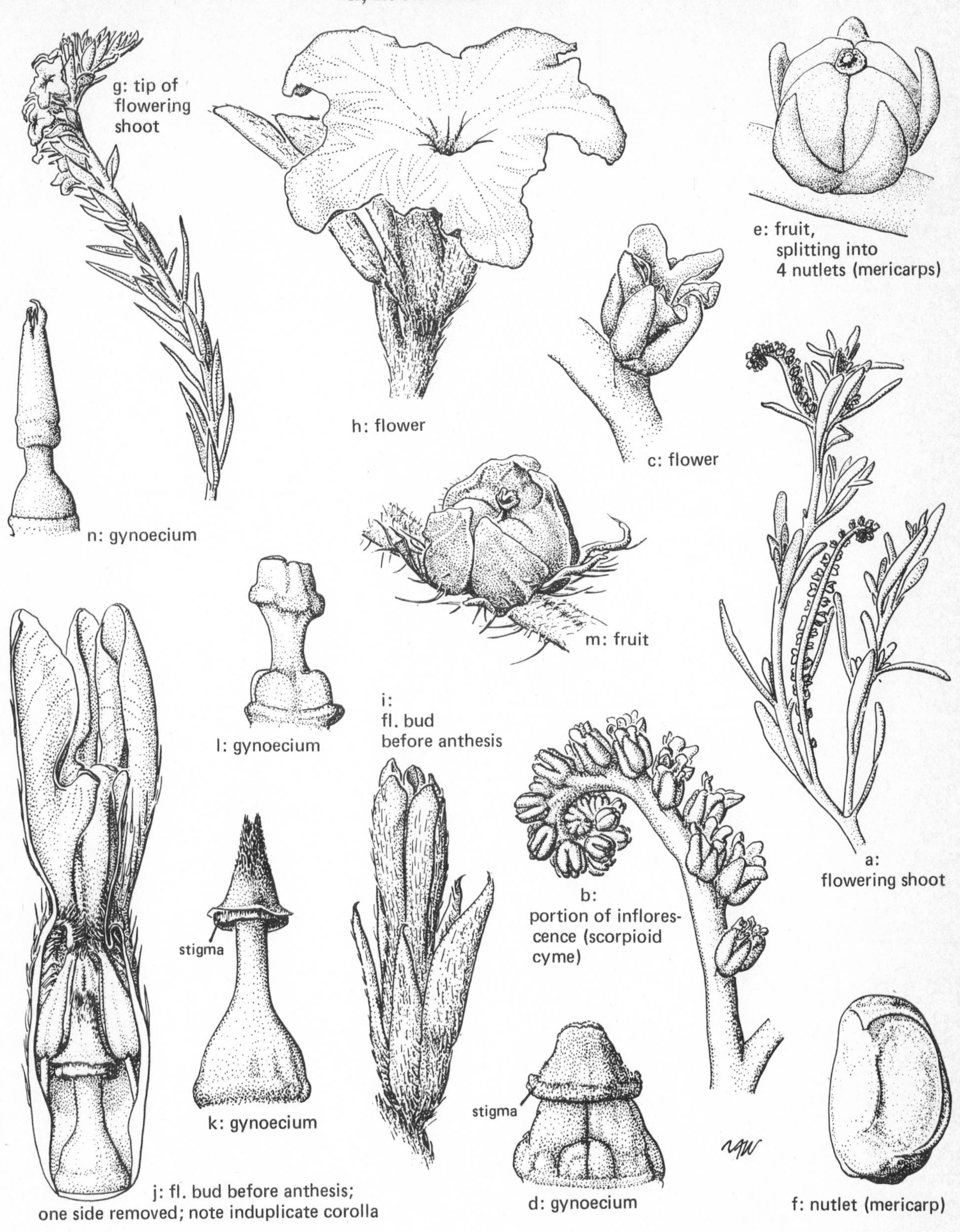

g: tip of flowering shoot

e: fruit, splitting into 4 nutlets (mericarps)

h: flower

c: flower

n: gynoecium

m: fruit

l: gynoecium

i: fl. bud before anthesis

a: flowering shoot

stigma

k: gynoecium

b: portion of inflorescence (scorpioid cyme)

stigma

d: gynoecium

j: fl. bud before anthesis; one side removed; note induplicate corolla

f: nutlet (mericarp)

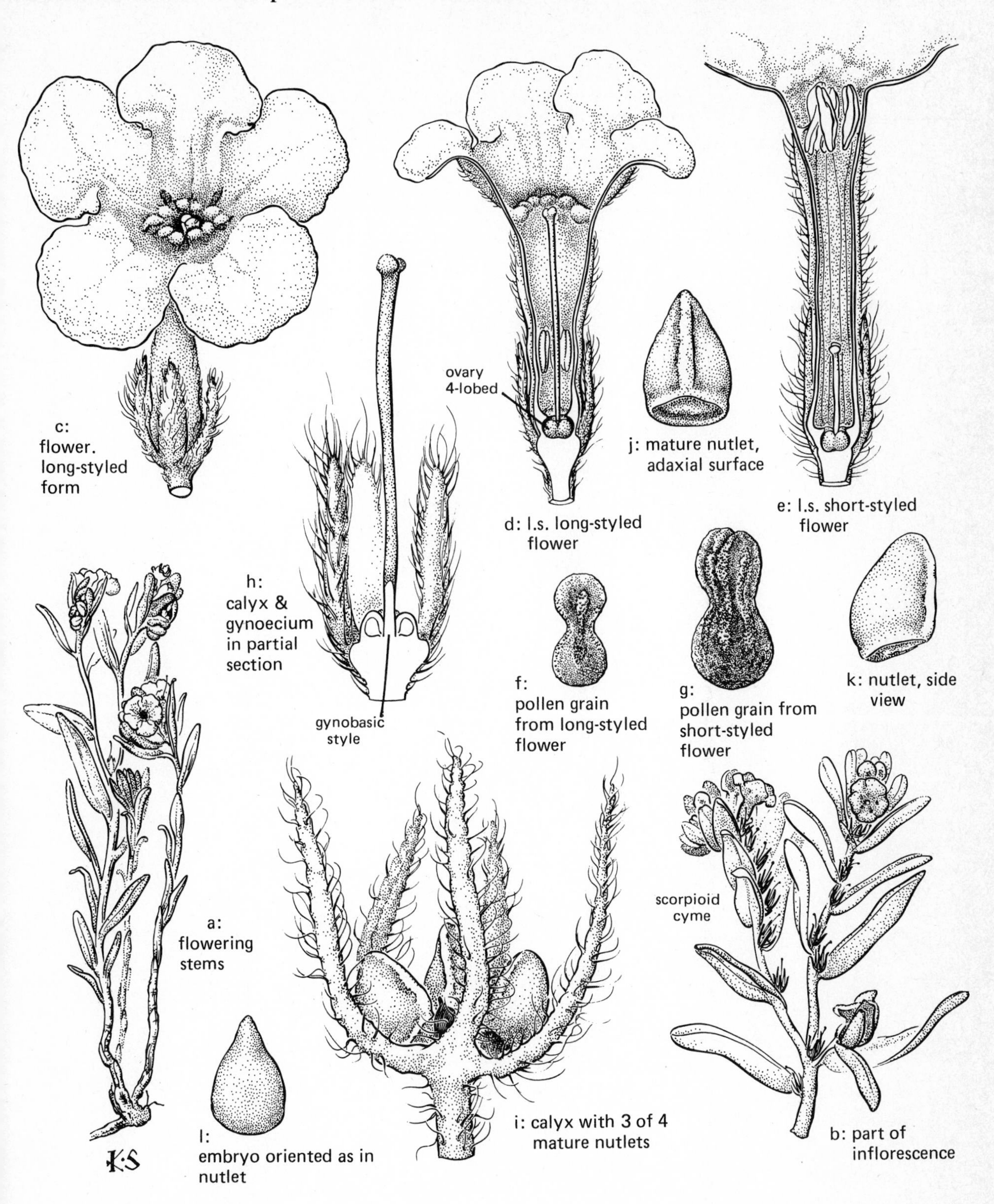

c:
flower.
long-styled
form

ovary
4-lobed

j: mature nutlet,
adaxial surface

d: l.s. long-styled
flower

e: l.s. short-styled
flower

h:
calyx &
gynoecium
in partial
section

gynobasic
style

f:
pollen grain
from long-styled
flower

g:
pollen grain from
short-styled
flower

k: nutlet, side
view

a:
flowering
stems

scorpioid
cyme

l:
embryo oriented as in
nutlet

i: calyx with 3 of 4
mature nutlets

b: part of
inflorescence

K·S

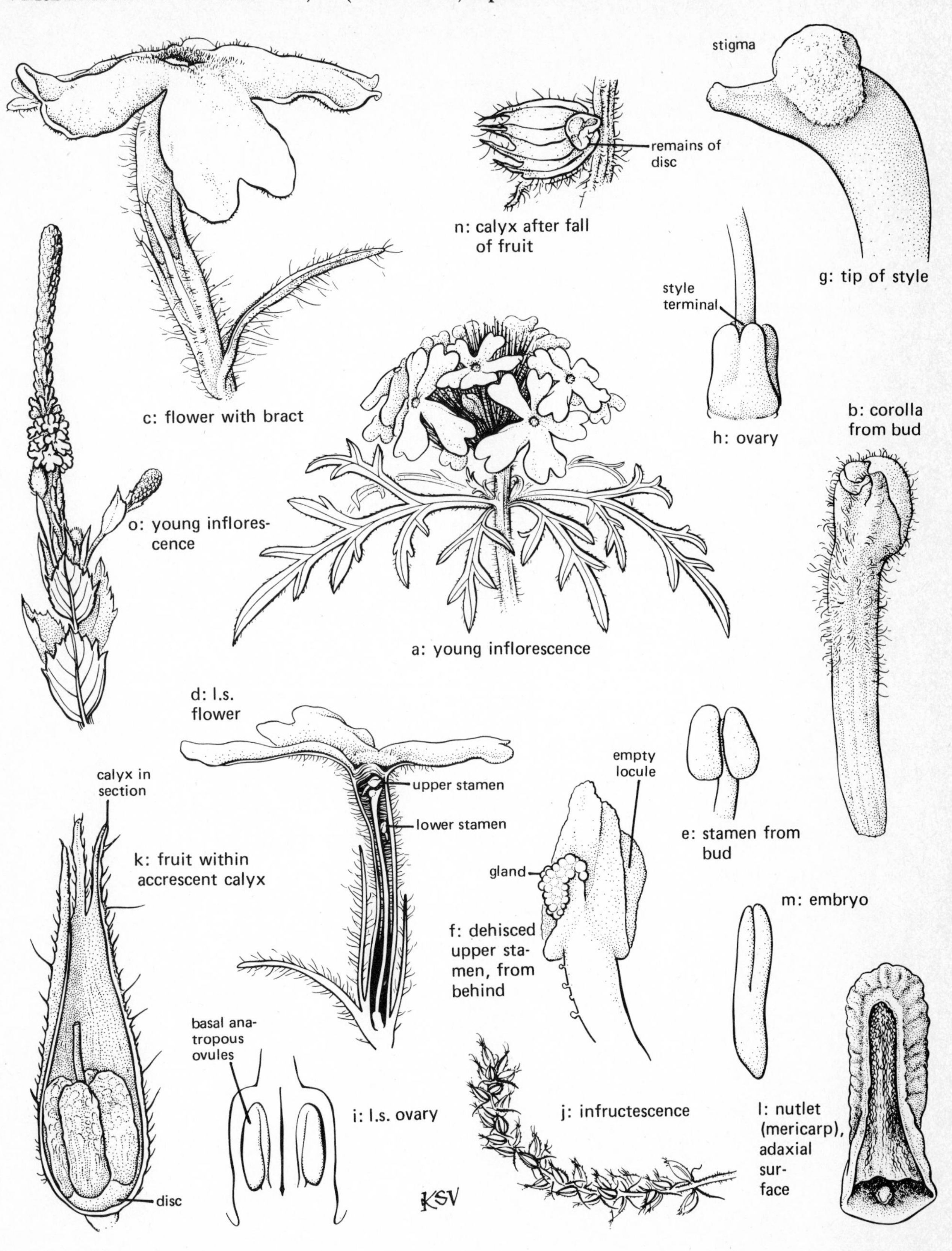

n: calyx after fall of fruit

remains of disc

stigma

g: tip of style

style terminal

h: ovary

b: corolla from bud

c: flower with bract

o: young inflorescence

a: young inflorescence

d: l.s. flower

upper stamen

lower stamen

calyx in section

k: fruit within accrescent calyx

gland

empty locule

e: stamen from bud

m: embryo

f: dehisced upper stamen, from behind

basal anatropous ovules

disc

i: l.s. ovary

j: infructescence

l: nutlet (mericarp), adaxial surface

KSV

LAMIACEAE (LABIATAE): Scutellaria. a-m, S. parvula; n-p, S. havanensis

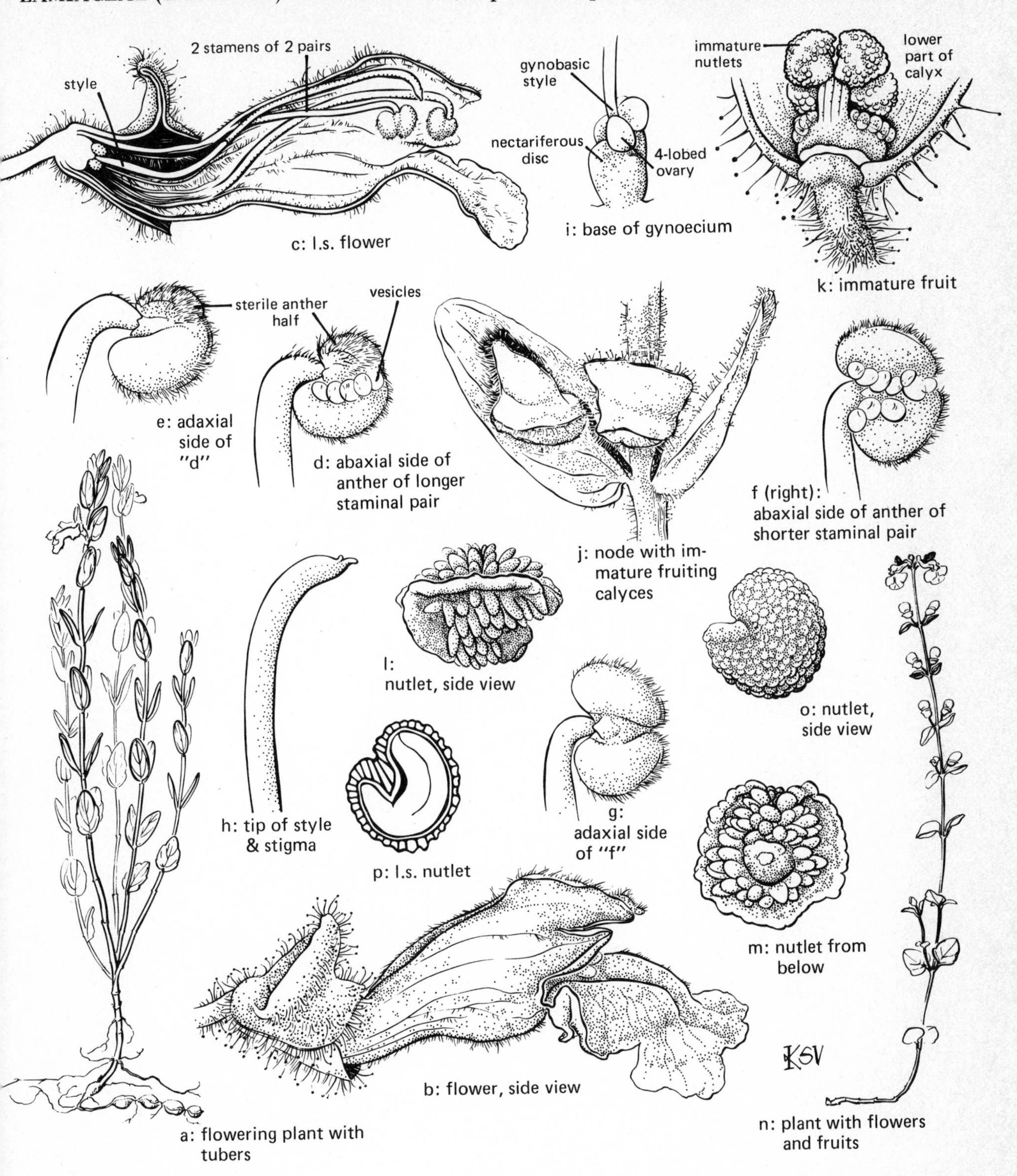

style

2 stamens of 2 pairs

c: l.s. flower

gynobasic style

nectariferous disc

4-lobed ovary

i: base of gynoecium

immature nutlets

lower part of calyx

k: immature fruit

sterile anther half

vesicles

e: adaxial side of "d"

d: abaxial side of anther of longer staminal pair

f (right): abaxial side of anther of shorter staminal pair

j: node with immature fruiting calyces

l: nutlet, side view

o: nutlet, side view

h: tip of style & stigma

p: l.s. nutlet

g: adaxial side of "f"

m: nutlet from below

b: flower, side view

a: flowering plant with tubers

n: plant with flowers and fruits

KSV

LAMIACEAE (LABIATAE): Salvia. a-j, S. urticifolia; k-o, S. lyrata

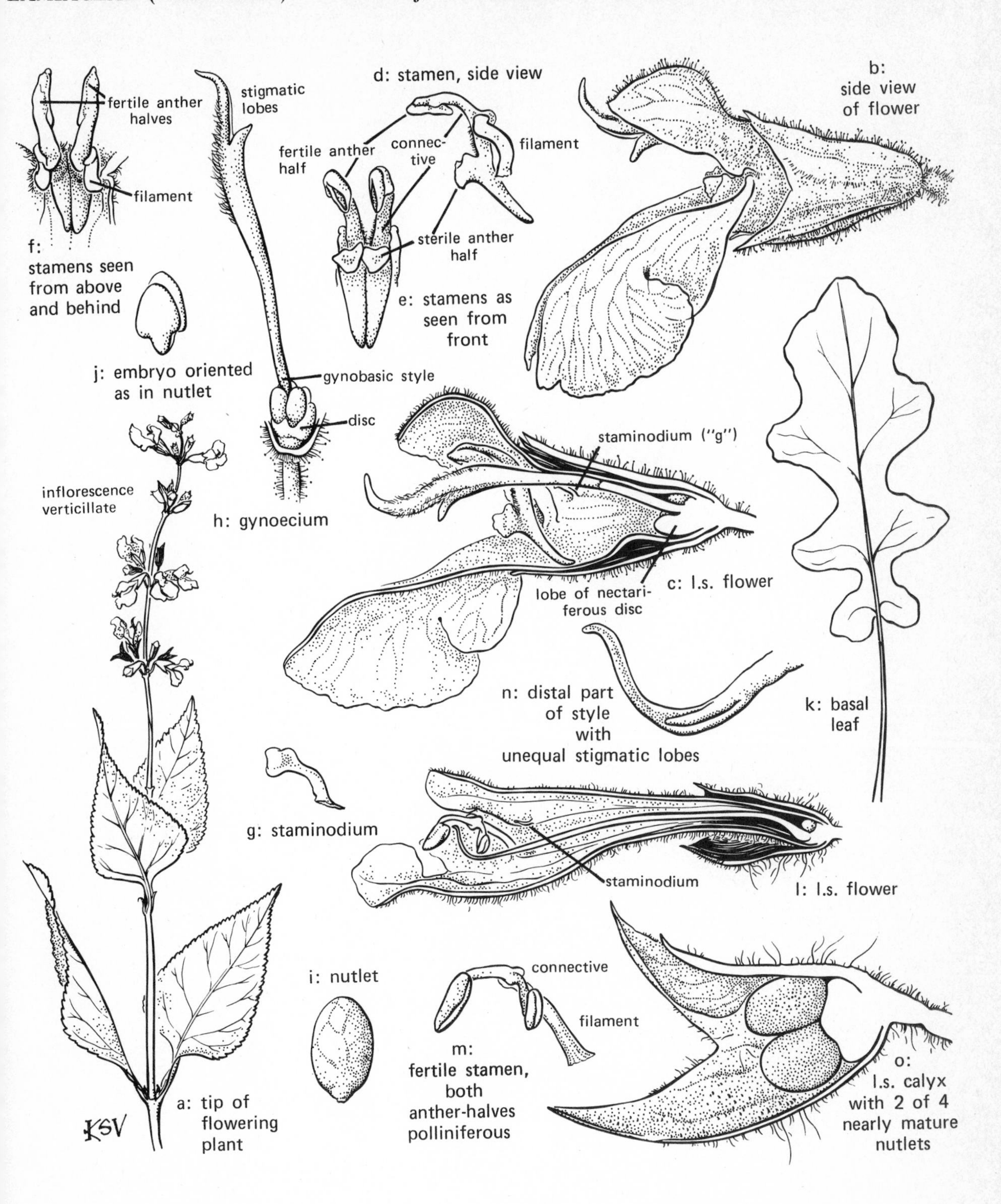

fertile anther halves

filament

f: stamens seen from above and behind

j: embryo oriented as in nutlet

inflorescence verticillate

a: tip of flowering plant

KSV

stigmatic lobes

d: stamen, side view

fertile anther half

connective

filament

sterile anther half

e: stamens as seen from front

gynobasic style

disc

h: gynoecium

g: staminodium

i: nutlet

m: fertile stamen, both anther-halves polliniferous

connective

filament

b: side view of flower

staminodium ("g")

lobe of nectari-ferous disc

c: l.s. flower

n: distal part of style with unequal stigmatic lobes

staminodium

l: l.s. flower

k: basal leaf

o: l.s. calyx with 2 of 4 nearly mature nutlets

SOLANACEAE: Physalis. a-j, P. heterophylla; k-m, P. viscosa var. maritima

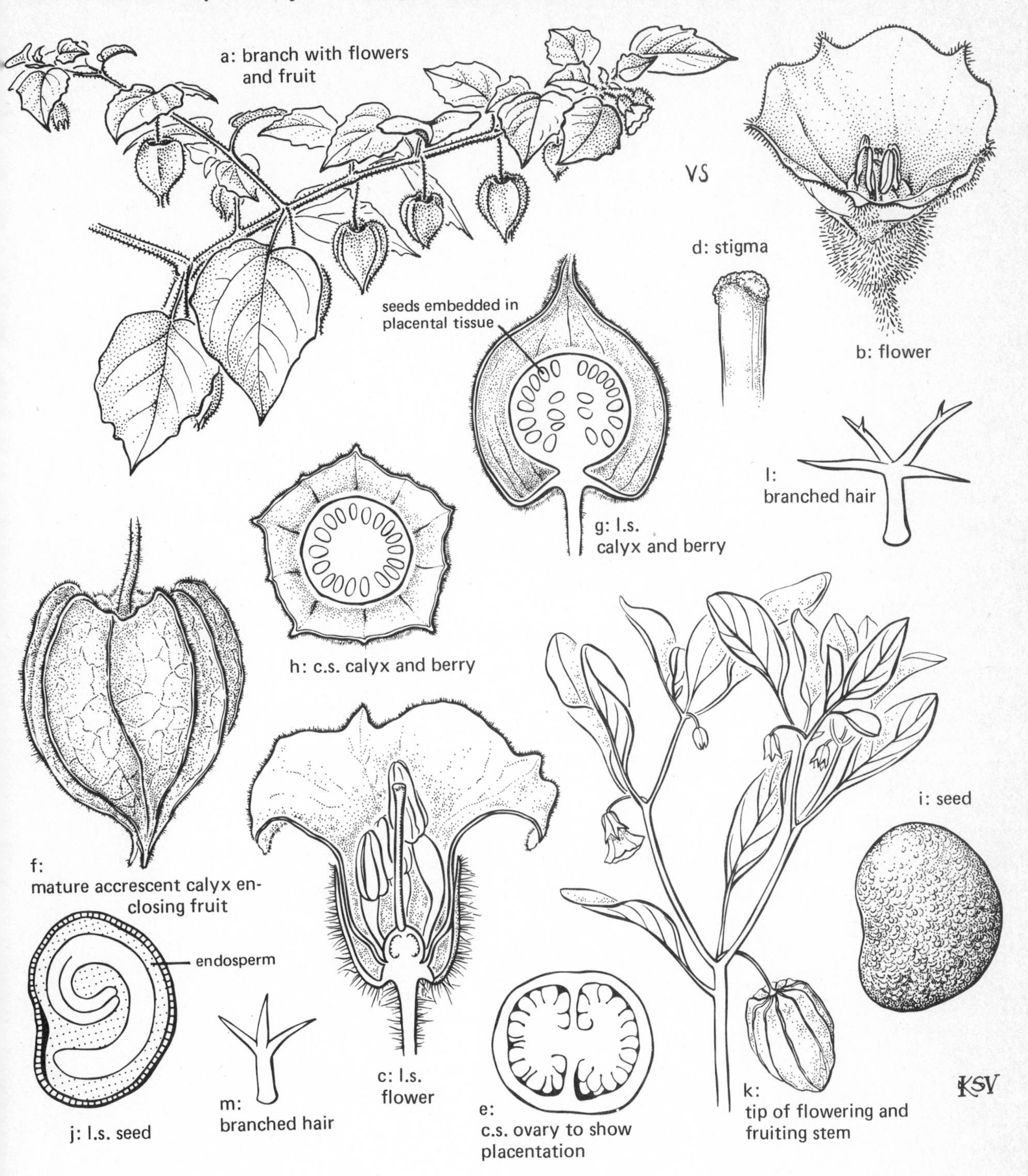

a: branch with flowers and fruit

seeds embedded in placental tissue

d: stigma

b: flower

VS

l: branched hair

g: l.s. calyx and berry

h: c.s. calyx and berry

i: seed

f: mature accrescent calyx enclosing fruit

endosperm

j: l.s. seed

m: branched hair

c: l.s. flower

e: c.s. ovary to show placentation

k: tip of flowering and fruiting stem

KSV

SCROPHULARIACEAE: Penstemon. a-j, P. canescens

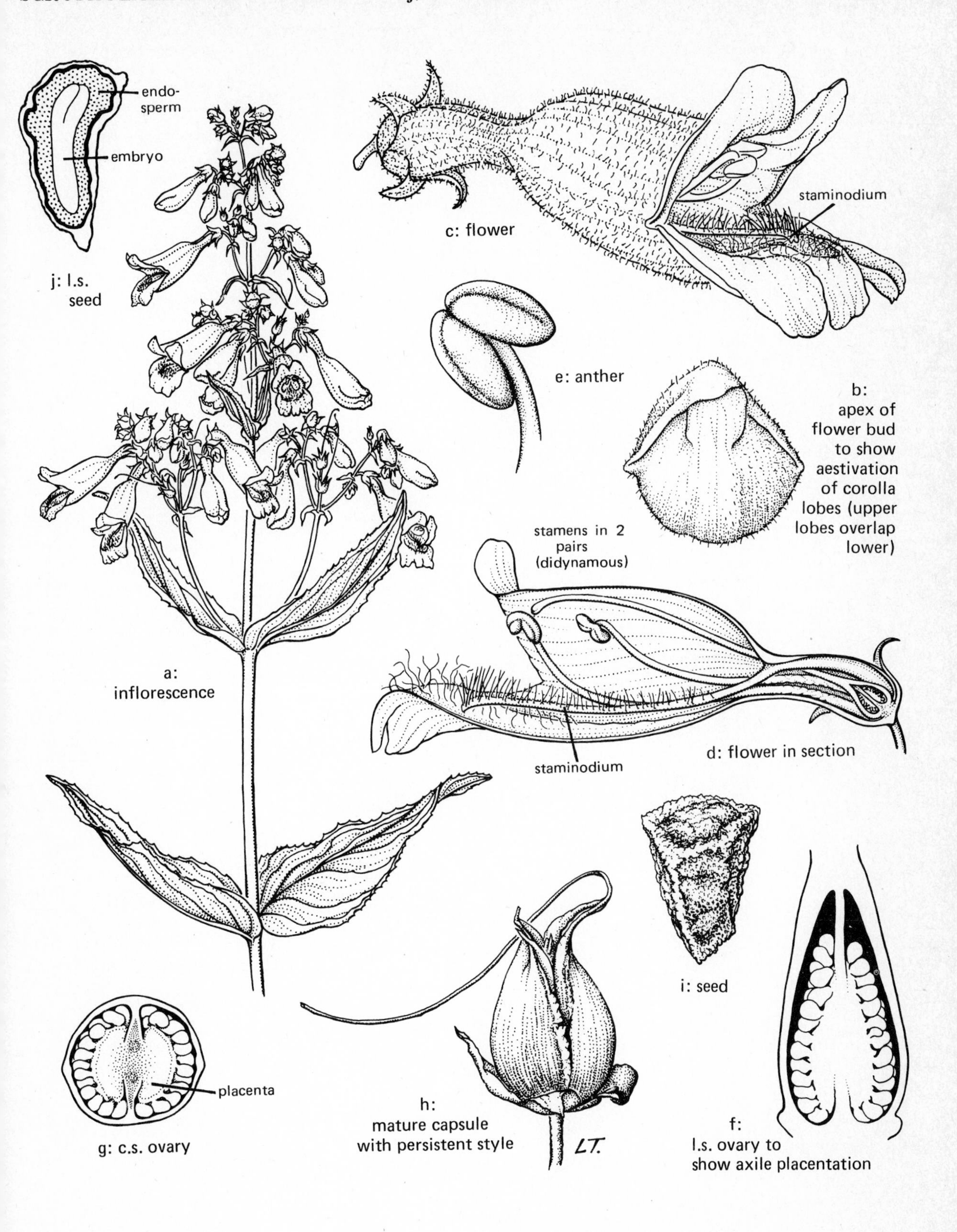

j: l.s. seed
endo-sperm
embryo

c: flower

staminodium

e: anther

b: apex of flower bud to show aestivation of corolla lobes (upper lobes overlap lower)

stamens in 2 pairs (didynamous)

a: inflorescence

staminodium

d: flower in section

i: seed

placenta

g: c.s. ovary

h: mature capsule with persistent style

LT.

f: l.s. ovary to show axile placentation

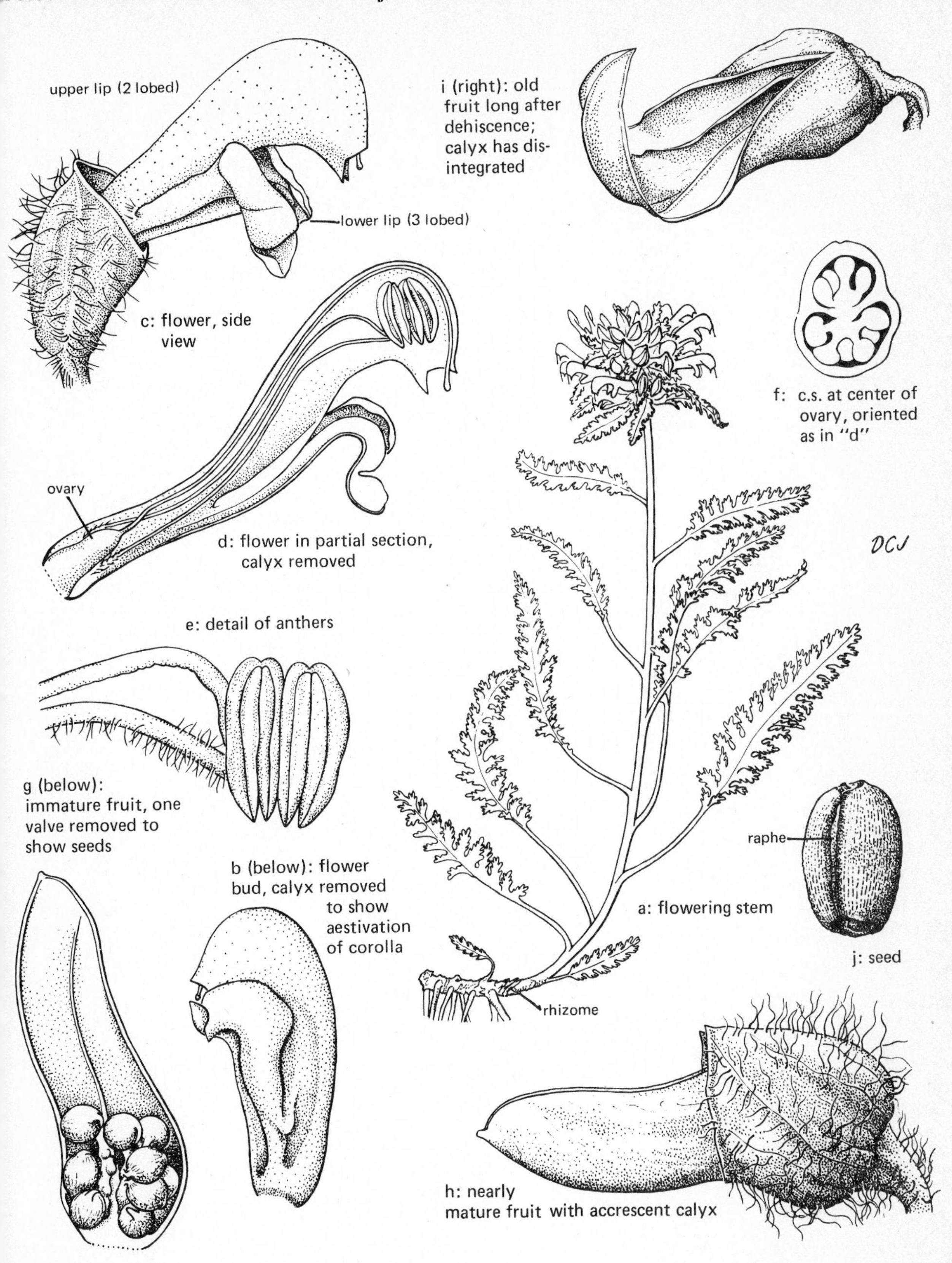

SCROPHULARIACEAE: Pedicularis. a-j, P. canadensis

upper lip (2 lobed)

i (right): old fruit long after dehiscence; calyx has disintegrated

lower lip (3 lobed)

c: flower, side view

f: c.s. at center of ovary, oriented as in "d"

ovary

d: flower in partial section, calyx removed

e: detail of anthers

g (below): immature fruit, one valve removed to show seeds

b (below): flower bud, calyx removed to show aestivation of corolla

a: flowering stem

raphe

rhizome

j: seed

h: nearly mature fruit with accrescent calyx

DCJ

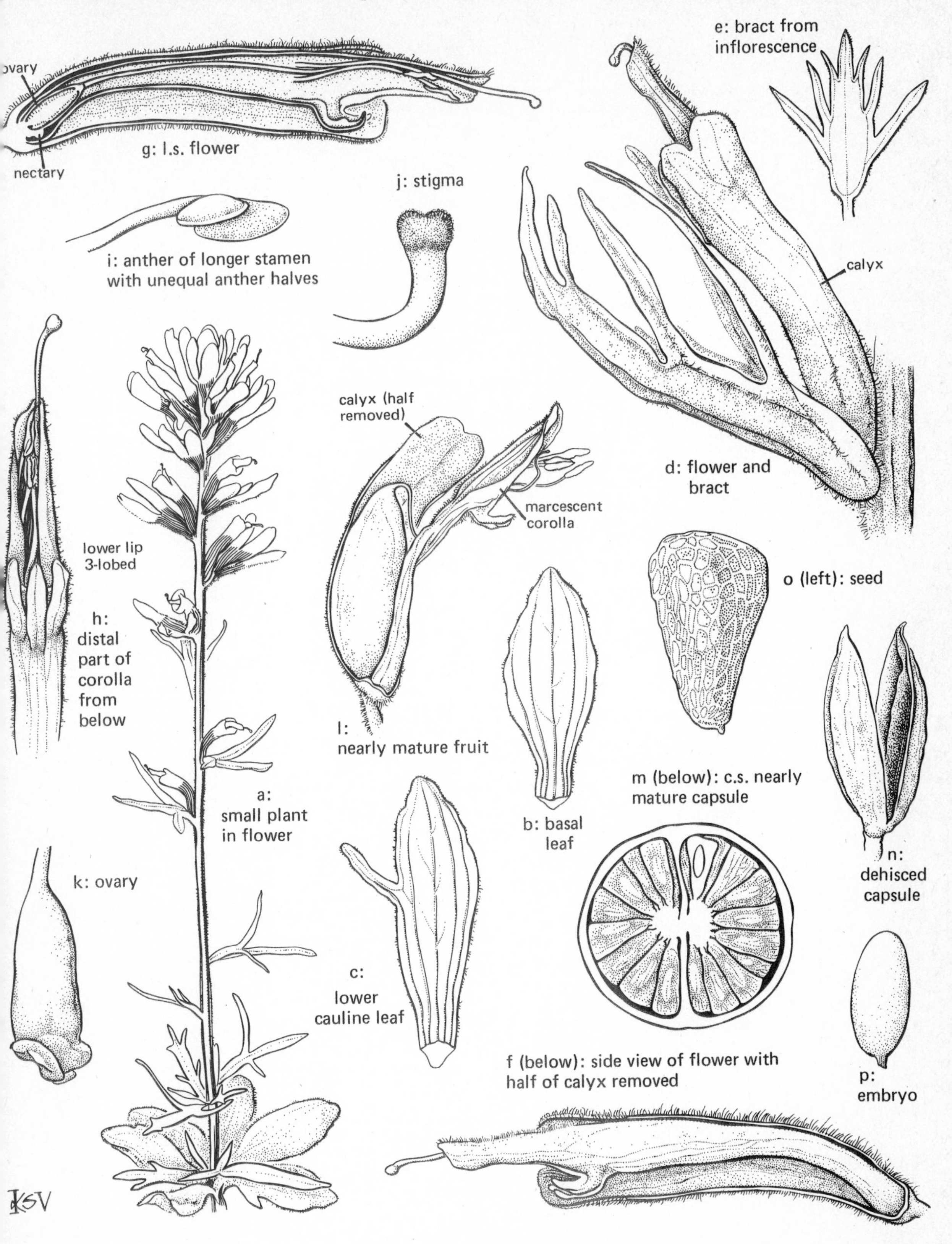

SCROPHULARIACEAE: Castilleia. a-p, C. coccinea

g: l.s. flower

ovary

nectary

e: bract from inflorescence

calyx

i: anther of longer stamen with unequal anther halves

j: stigma

calyx (half removed)

marcescent corolla

d: flower and bract

lower lip 3-lobed

h: distal part of corolla from below

a: small plant in flower

l: nearly mature fruit

b: basal leaf

o (left): seed

m (below): c.s. nearly mature capsule

n: dehisced capsule

k: ovary

c: lower cauline leaf

f (below): side view of flower with half of calyx removed

p: embryo

KSV

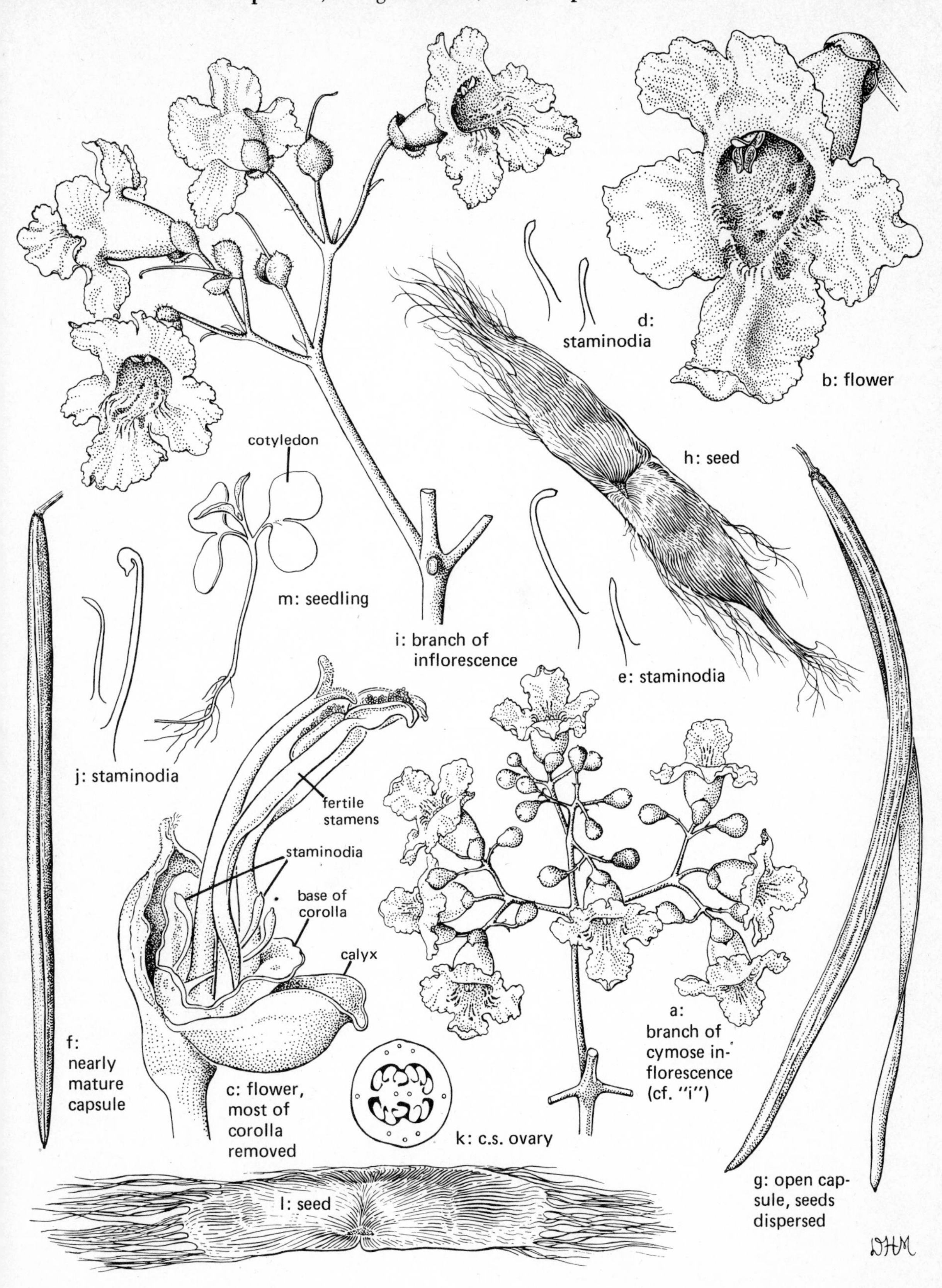

d: staminodia

b: flower

h: seed

cotyledon

m: seedling

i: branch of inflorescence

e: staminodia

j: staminodia

fertile stamens

staminodia

base of corolla

calyx

a: branch of cymose in-florescence (cf. "i")

f: nearly mature capsule

c: flower, most of corolla removed

k: c.s. ovary

l: seed

g: open cap-sule, seeds dispersed

DHM

ROBANCHACEAE: Orobanche. a-k, O. uniflora

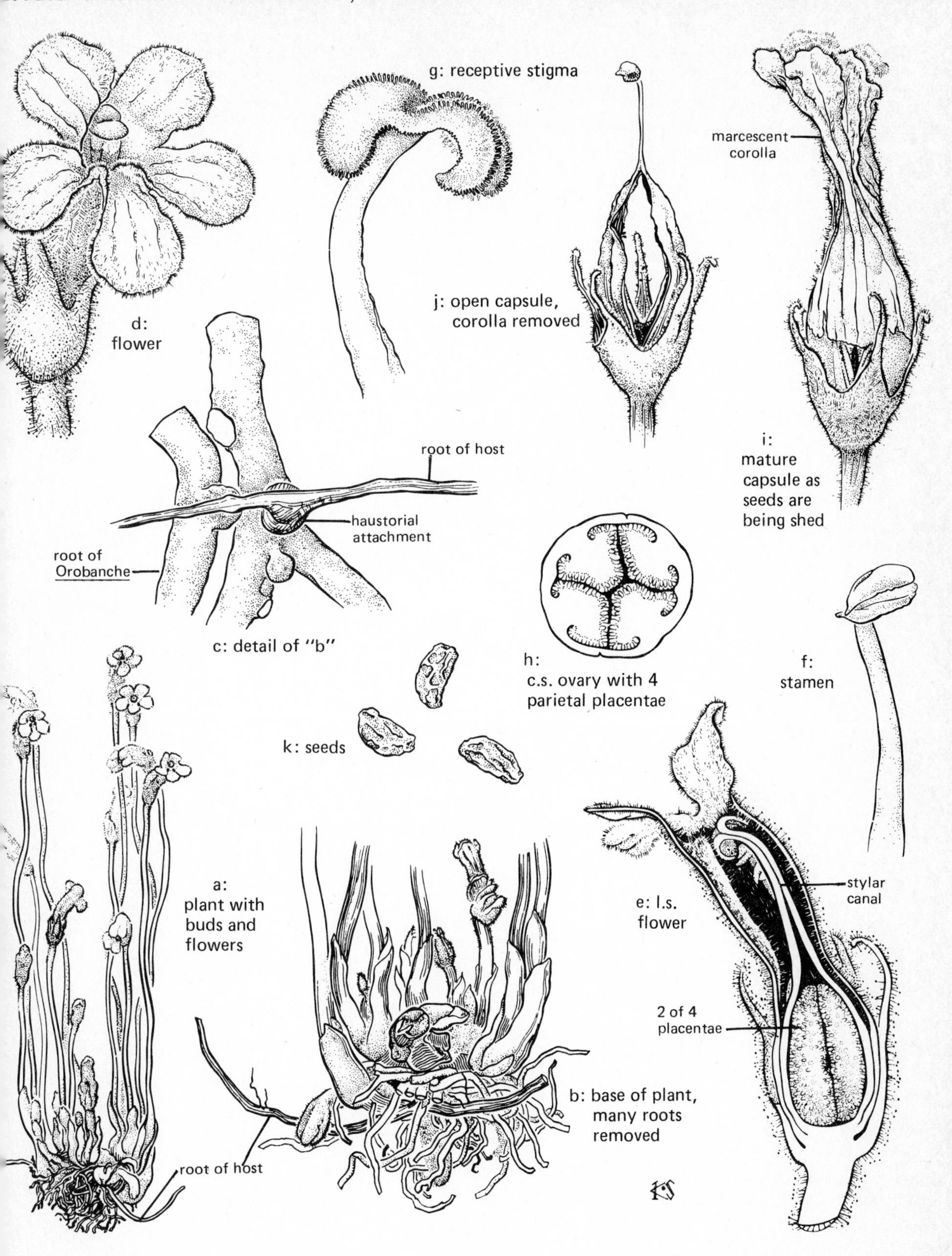

d: flower

g: receptive stigma

j: open capsule, corolla removed

marcescent corolla

i: mature capsule as seeds are being shed

root of host

haustorial attachment

root of Orobanche

c: detail of "b"

h: c.s. ovary with 4 parietal placentae

f: stamen

k: seeds

a: plant with buds and flowers

e: l.s. flower

stylar canal

2 of 4 placentae

root of host

b: base of plant, many roots removed

ACANTHACEAE: Justicia. a-g, J. ovata var. angusta; h, J. crassifolia; i-l, J. cooleyi; m, n, J. americana

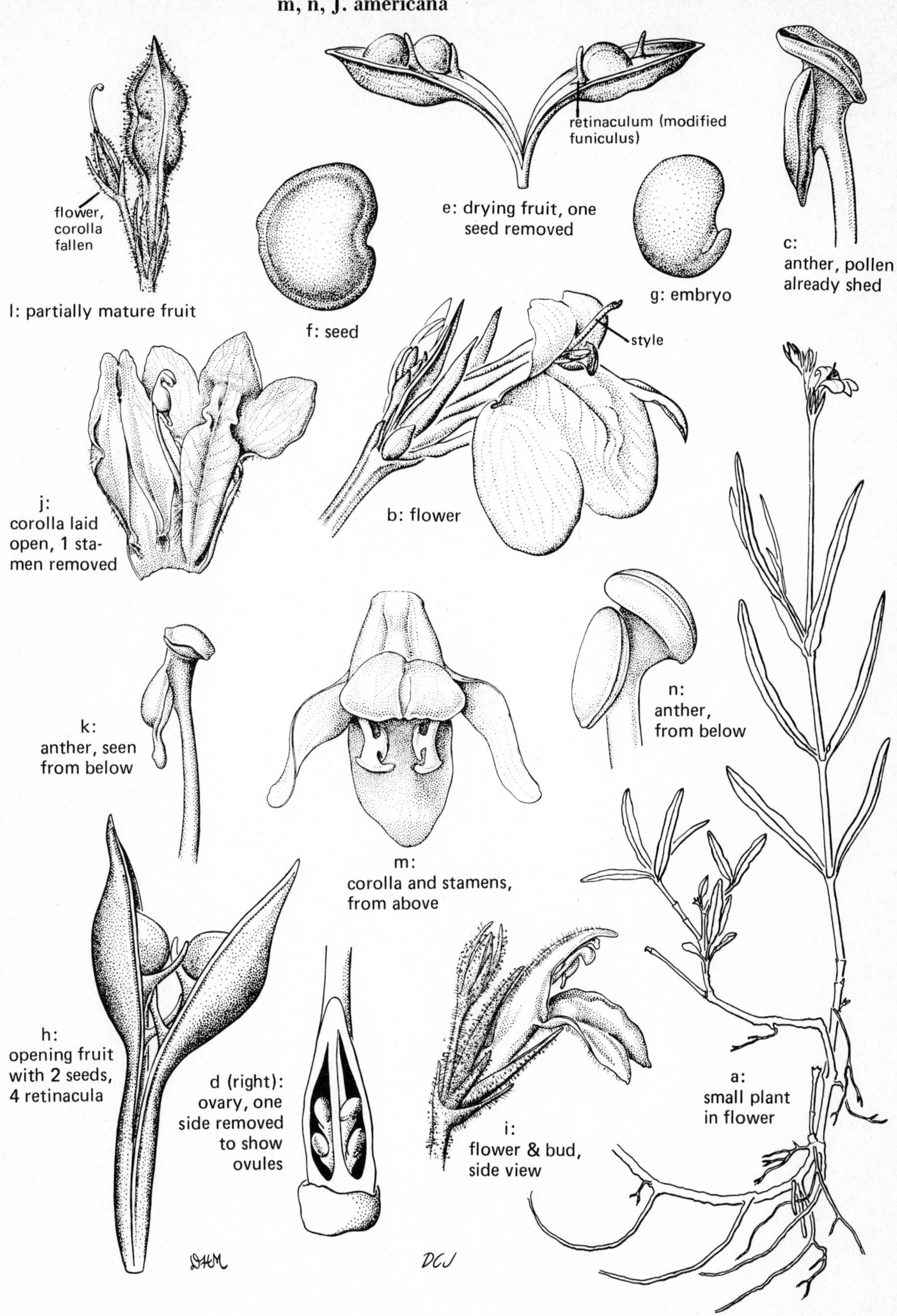

flower, corolla fallen

l: partially mature fruit

retinaculum (modified funiculus)

e: drying fruit, one seed removed

f: seed

g: embryo

c: anther, pollen already shed

style

b: flower

j: corolla laid open, 1 stamen removed

k: anther, seen from below

n: anther, from below

m: corolla and stamens, from above

h: opening fruit with 2 seeds, 4 retinacula

d (right): ovary, one side removed to show ovules

i: flower & bud, side view

a: small plant in flower

PLANTAGINACEAE: Plantago. a-i, P. lanceolata; j-l, P. aristata; m, n, P. rugelii; o, p, P. major; q, P. virginica

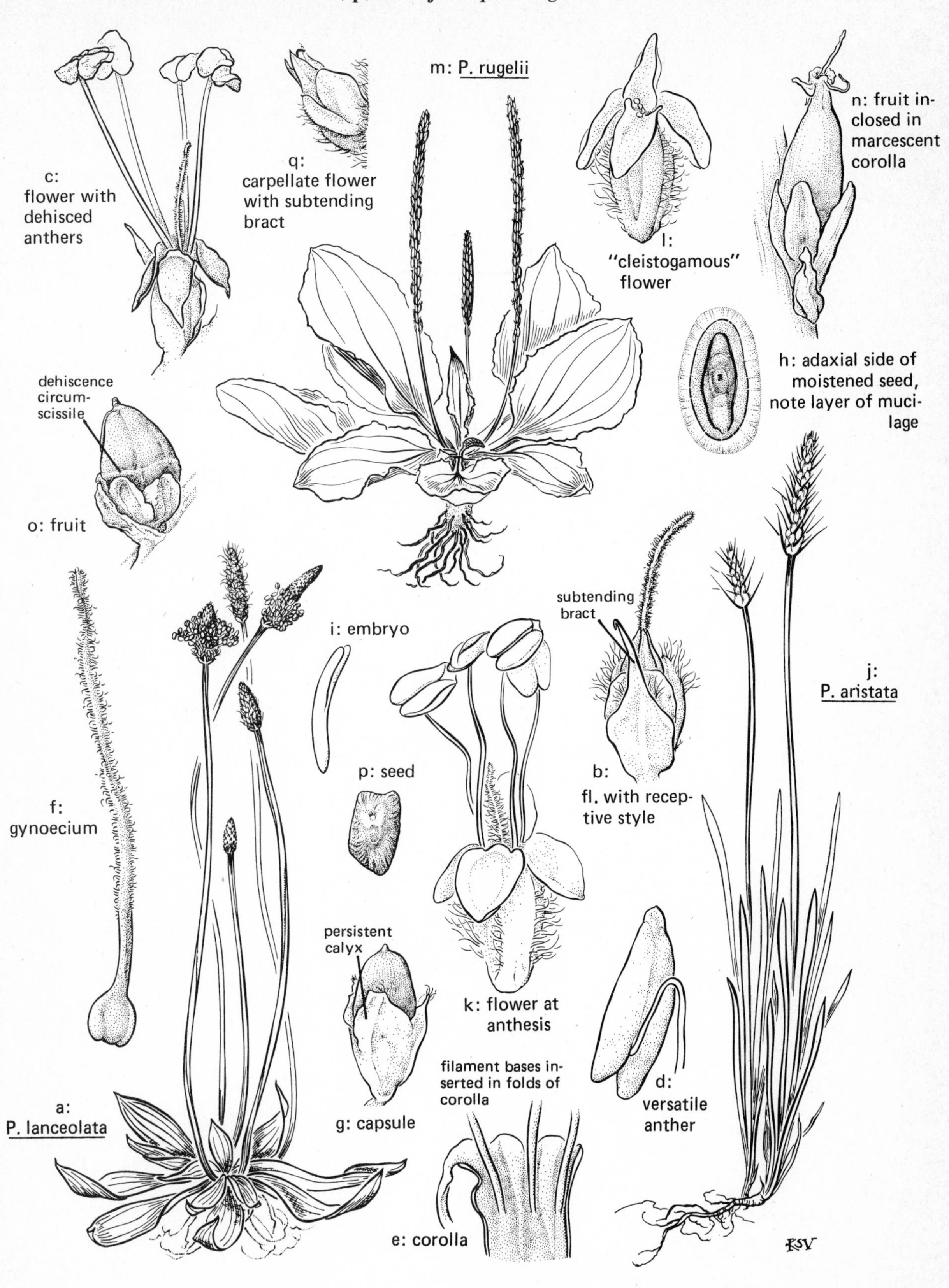

m: P. rugelii

c: flower with dehisced anthers

q: carpellate flower with subtending bract

n: fruit inclosed in marcescent corolla

l: "cleistogamous" flower

h: adaxial side of moistened seed, note layer of mucilage

dehiscence circumscissile

o: fruit

i: embryo

subtending bract

j: P. aristata

p: seed

b: fl. with receptive style

f: gynoecium

persistent calyx

k: flower at anthesis

filament bases inserted in folds of corolla

d: versatile anther

a: P. lanceolata

g: capsule

e: corolla

RSV

105

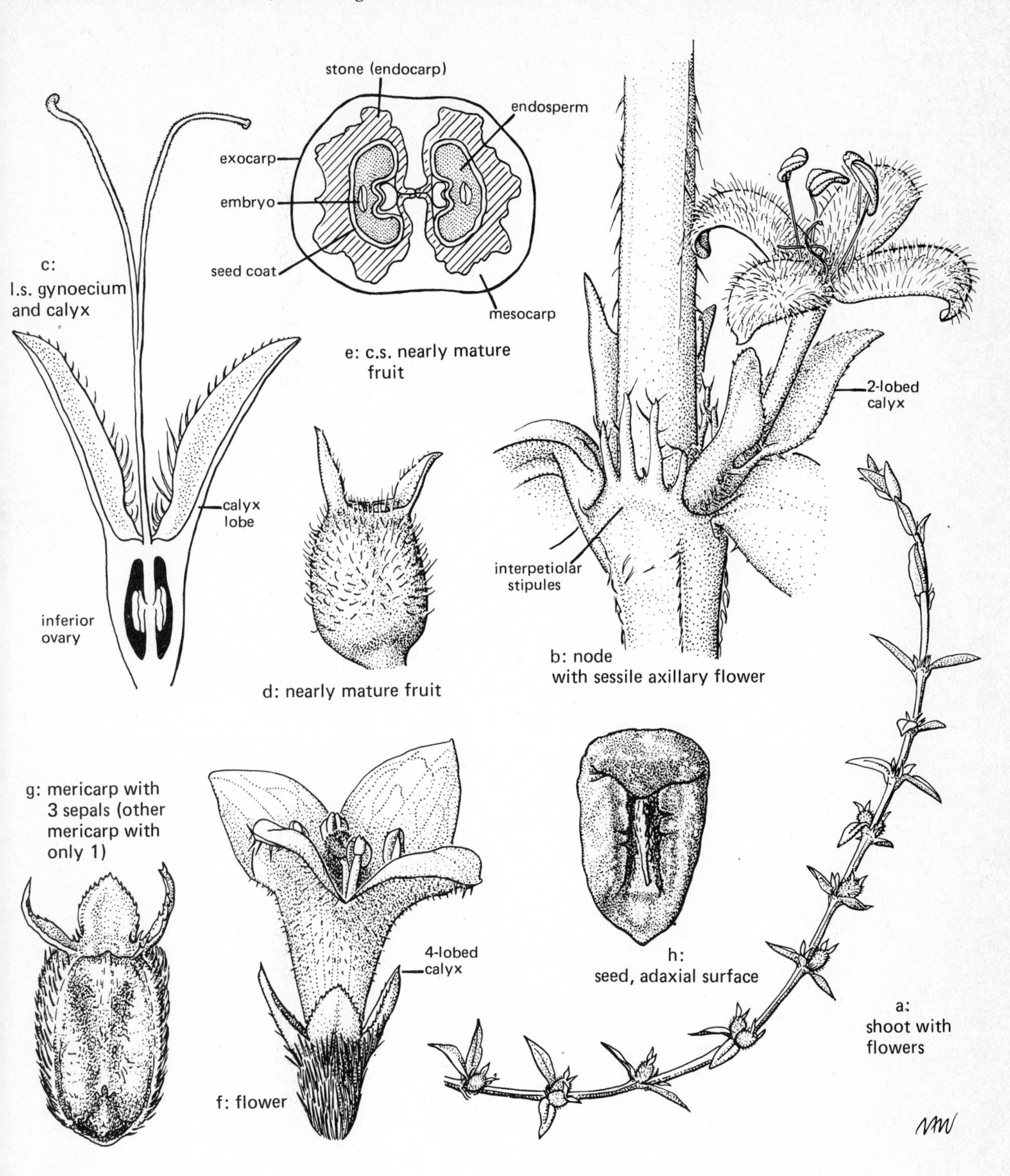

c:
l.s. gynoecium
and calyx

calyx
lobe

inferior
ovary

stone (endocarp)

exocarp

embryo

seed coat

endosperm

mesocarp

e: c.s. nearly mature
fruit

d: nearly mature fruit

2-lobed
calyx

interpetiolar
stipules

b: node
with sessile axillary flower

g: mericarp with
3 sepals (other
mericarp with
only 1)

4-lobed
calyx

f: flower

h:
seed, adaxial surface

a:
shoot with
flowers

RUBIACEAE: Mitchella. a-m, M. repens

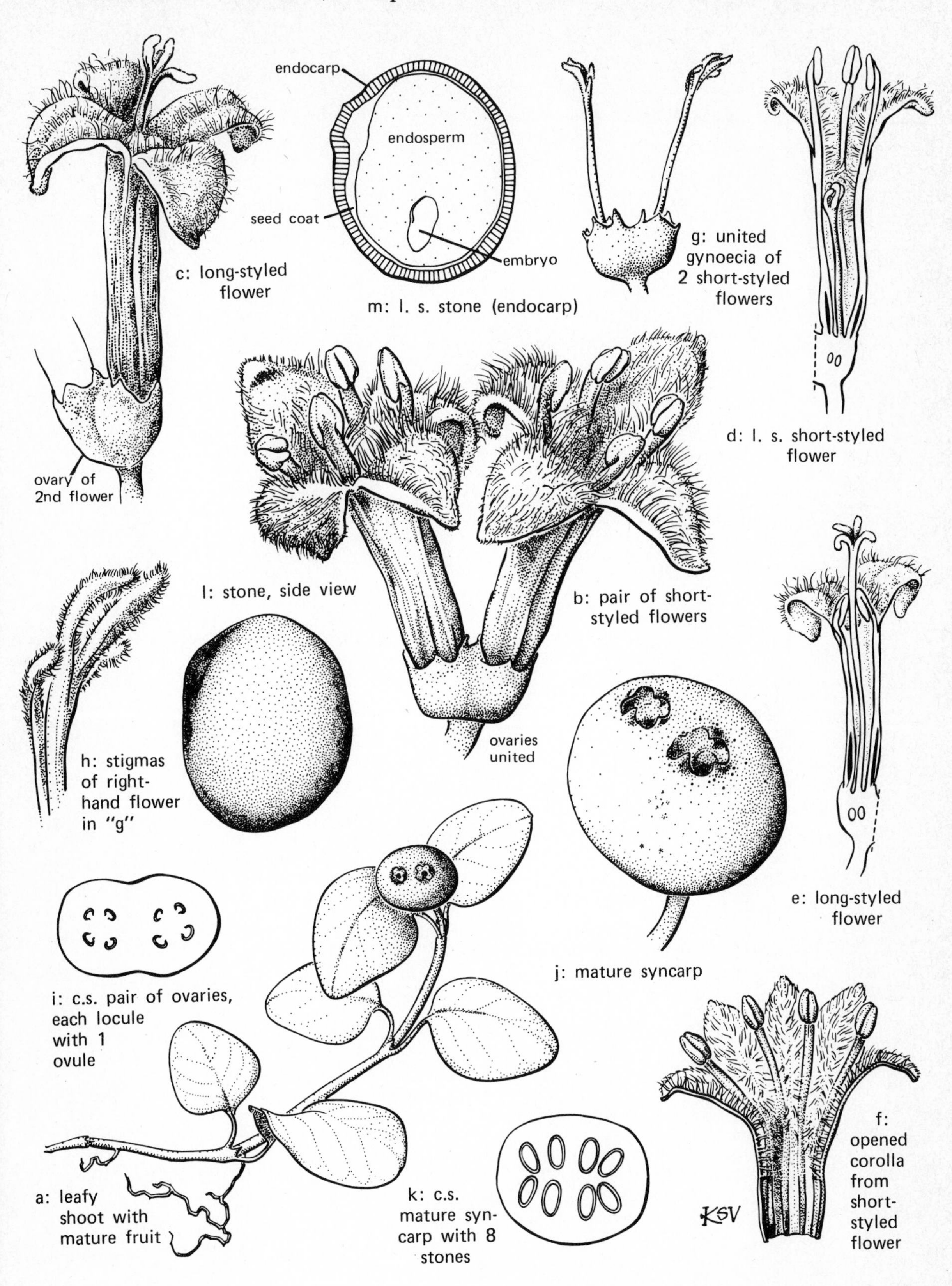

c: long-styled flower

ovary of 2nd flower

endocarp

endosperm

seed coat

embryo

m: l. s. stone (endocarp)

g: united gynoecia of 2 short-styled flowers

d: l. s. short-styled flower

h: stigmas of right-hand flower in "g"

l: stone, side view

b: pair of short-styled flowers

ovaries united

e: long-styled flower

i: c.s. pair of ovaries, each locule with 1 ovule

j: mature syncarp

a: leafy shoot with mature fruit

k: c.s. mature syncarp with 8 stones

f: opened corolla from short-styled flower

KSV

107

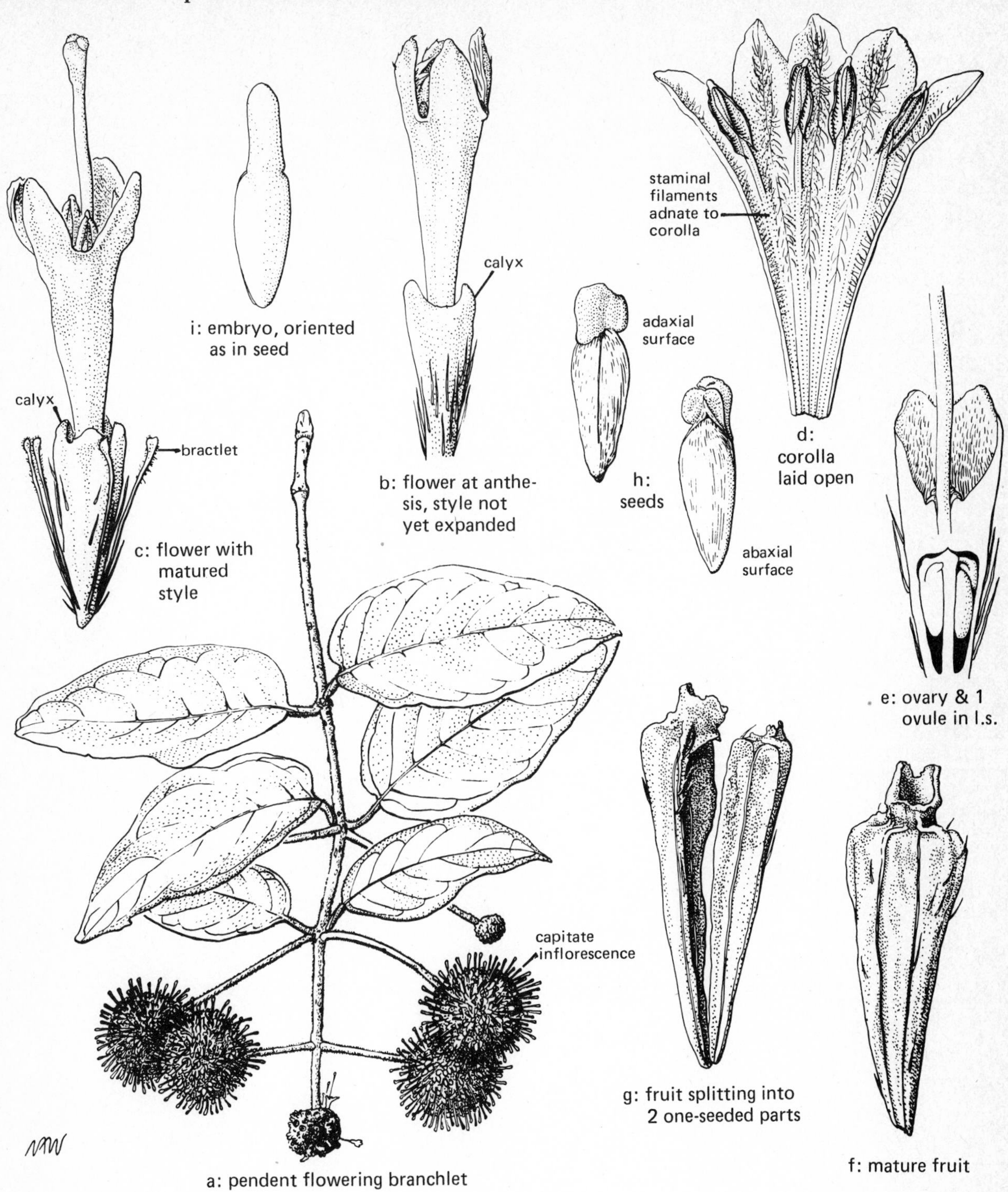

i: embryo, oriented
as in seed

calyx

bractlet

c: flower with
matured
style

calyx

b: flower at anthe-
sis, style not
yet expanded

h:
seeds

adaxial
surface

abaxial
surface

staminal
filaments
adnate to
corolla

d:
corolla
laid open

e: ovary & 1
ovule in l.s.

capitate
inflorescence

g: fruit splitting into
2 one-seeded parts

a: pendent flowering branchlet

f: mature fruit

CAPRIFOLIACEAE: Sambucus. a-f, *S. canadensis*

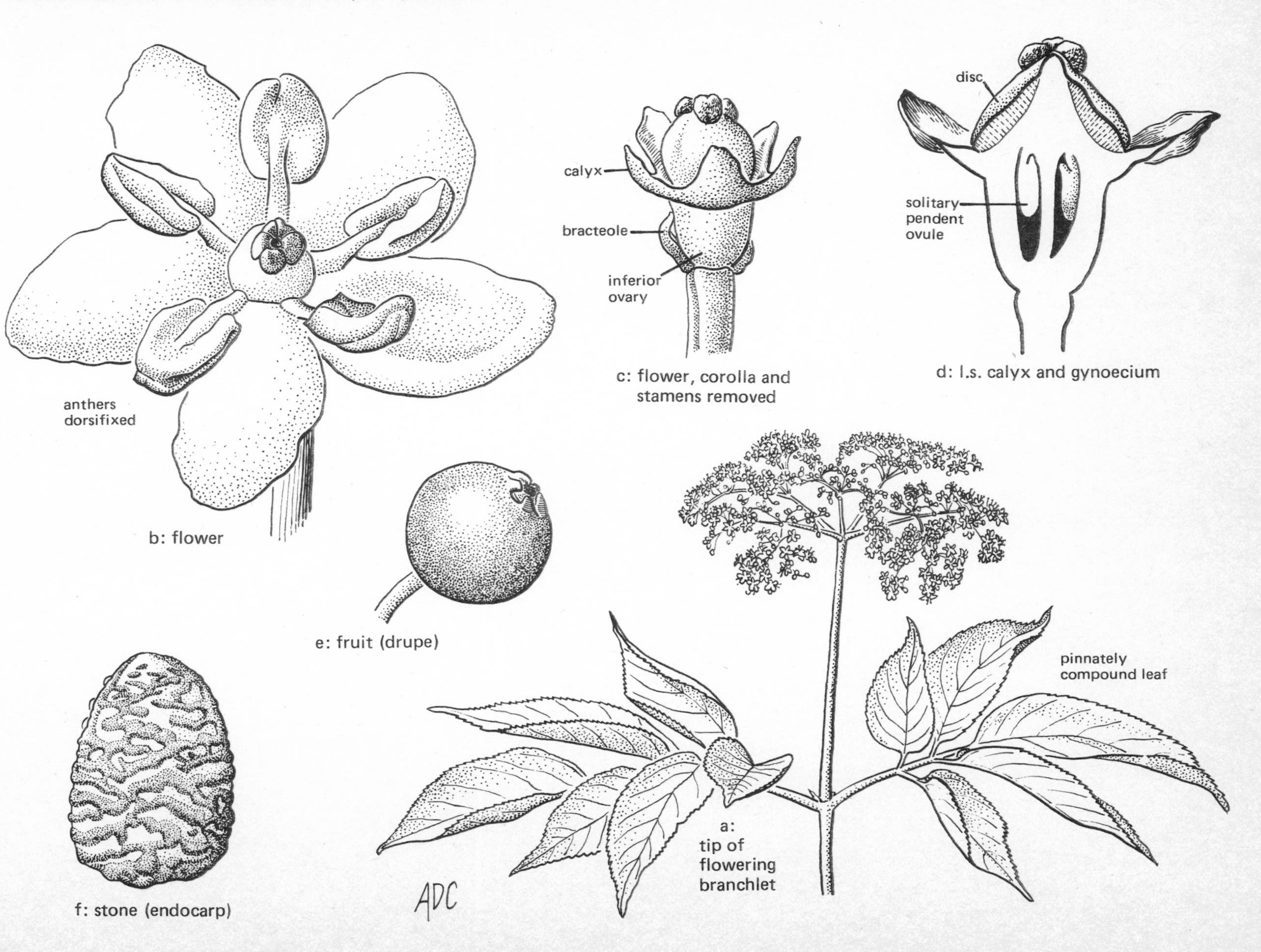

disc

solitary
pendent
ovule

d: l.s. calyx and gynoecium

calyx

bracteole

inferior
ovary

c: flower, corolla and
stamens removed

pinnately
compound leaf

a:
tip of
flowering
branchlet

APC

anthers
dorsifixed

b: flower

e: fruit (drupe)

f: stone (endocarp)

CAPRIFOLIACEAE: Lonicera. a-h, L. sempervirens; i-j, L. japonica

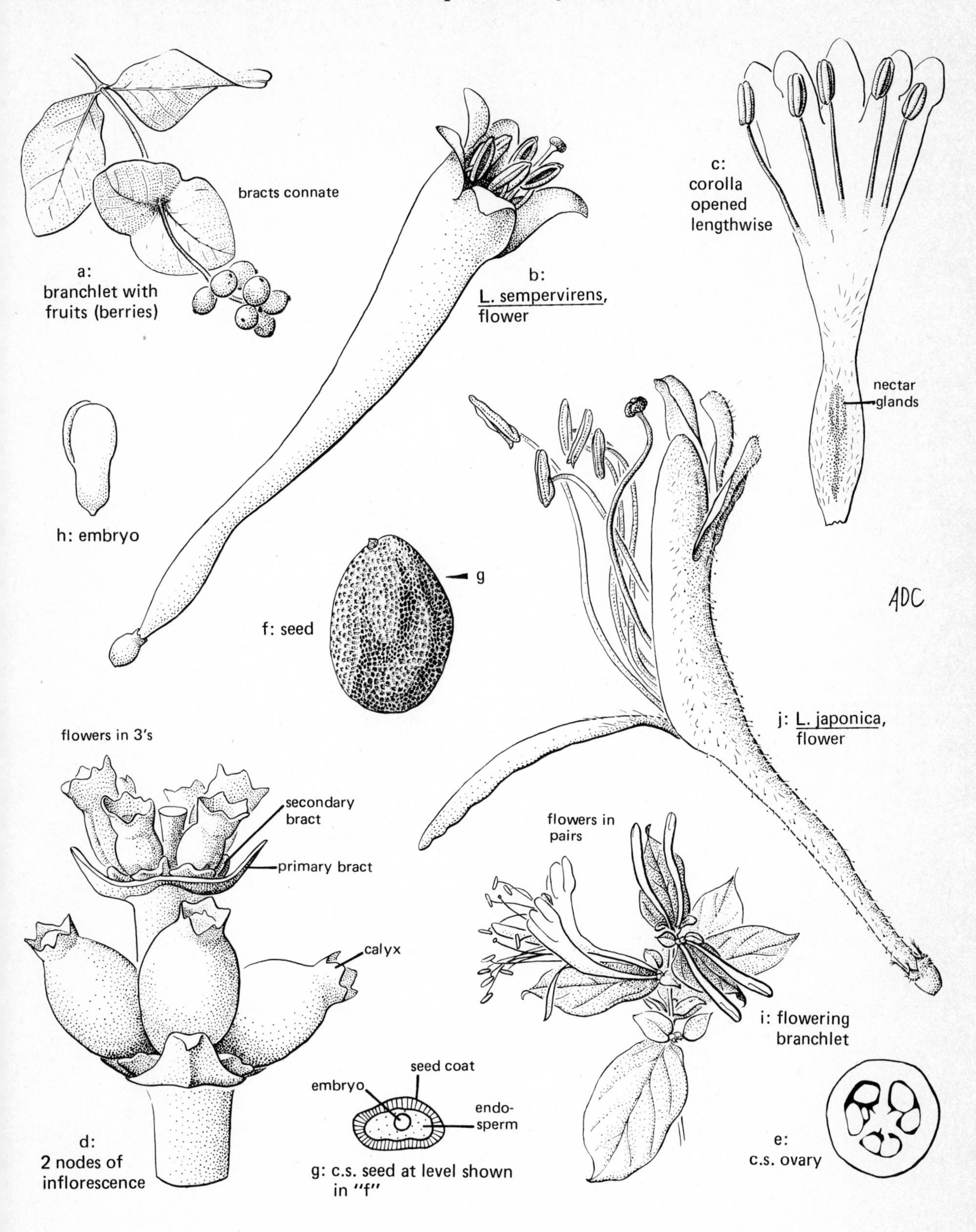

bracts connate

a:
branchlet with
fruits (berries)

b:
L. sempervirens,
flower

c:
corolla
opened
lengthwise

nectar
glands

h: embryo

g

f: seed

ADC

j: L. japonica,
flower

flowers in 3's

secondary
bract

primary
bract

calyx

flowers in
pairs

i: flowering
branchlet

d:
2 nodes of
inflorescence

seed coat

embryo

endo-
sperm

g: c.s. seed at level shown
in "f"

e:
c.s. ovary

CUCURBITACEAE: Echinocystis. a-i, E. lobatus

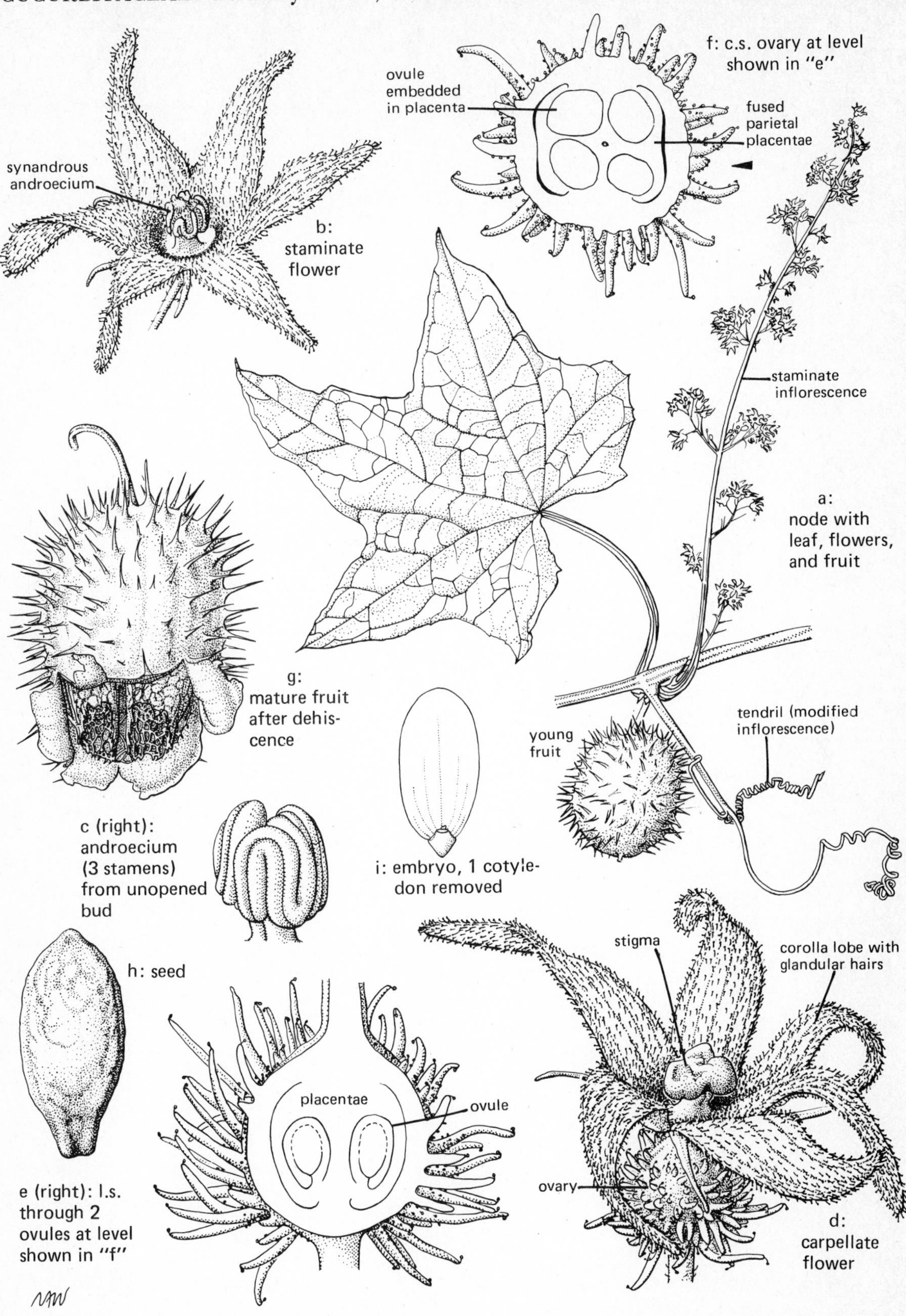

synandrous androecium

b: staminate flower

f: c.s. ovary at level shown in "e"

ovule embedded in placenta

fused parietal placentae

staminate inflorescence

a: node with leaf, flowers, and fruit

g: mature fruit after dehiscence

young fruit

tendril (modified inflorescence)

c (right): androecium (3 stamens) from unopened bud

i: embryo, 1 cotyledon removed

h: seed

stigma

corolla lobe with glandular hairs

placentae

ovule

e (right): l.s. through 2 ovules at level shown in "f"

ovary

d: carpellate flower

CUCURBITACEAE: Sicyos. a–h, S. angulatus

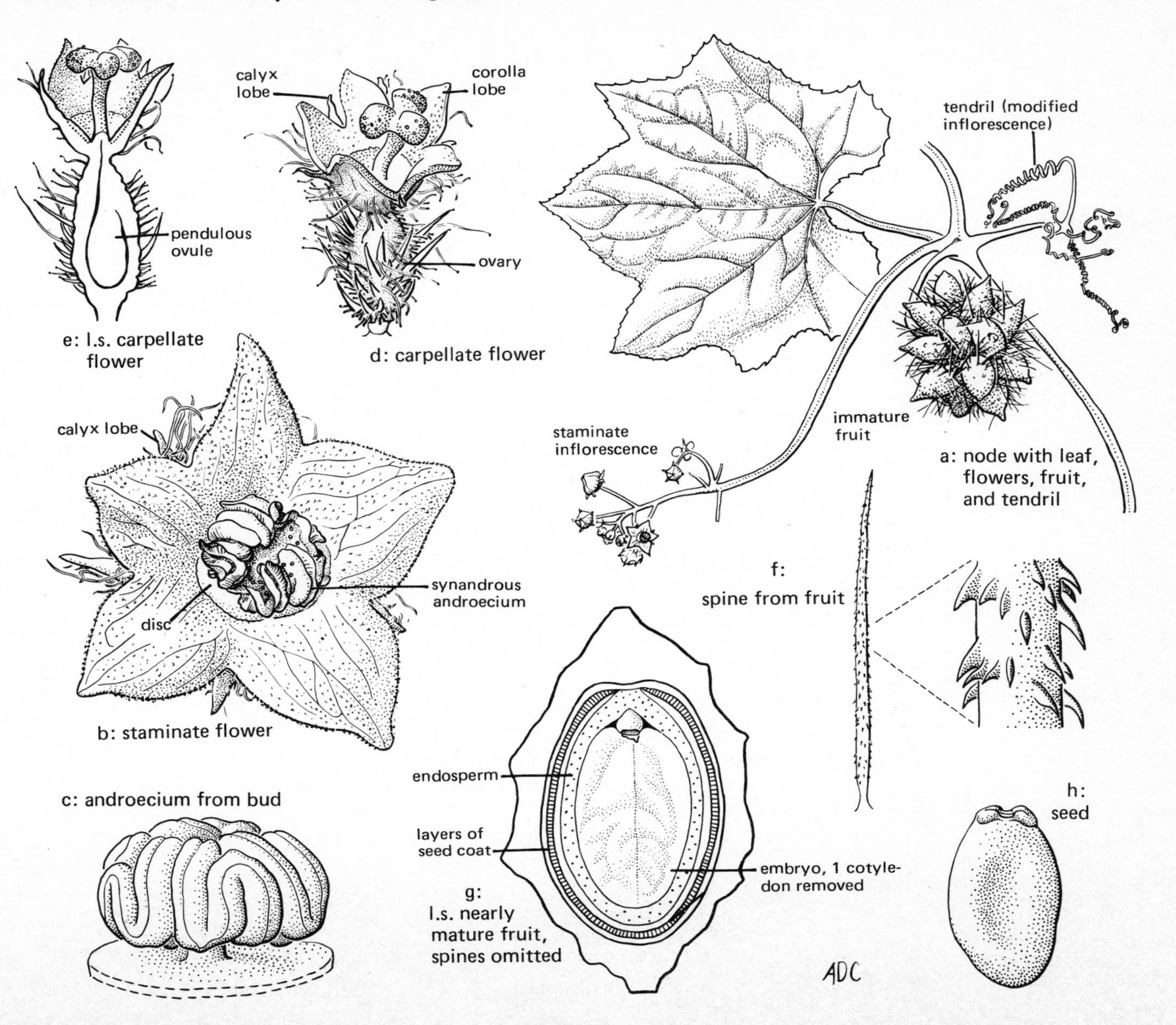

tendril (modified inflorescence)

immature fruit

a: node with leaf, flowers, fruit, and tendril

staminate inflorescence

f: spine from fruit

h: seed

corolla lobe

calyx lobe

ovary

d: carpellate flower

pendulous ovule

e: l.s. carpellate flower

calyx lobe

disc

synandrous androoecium

b: staminate flower

endosperm

layers of seed coat

embryo, 1 cotyledon removed

g: l.s. nearly mature fruit, spines omitted

c: androecium from bud

ADC

CAMPANULACEAE subfamily CAMPANULOIDEAE: Triodanis. a-g, T. (Specularia) perfoliata

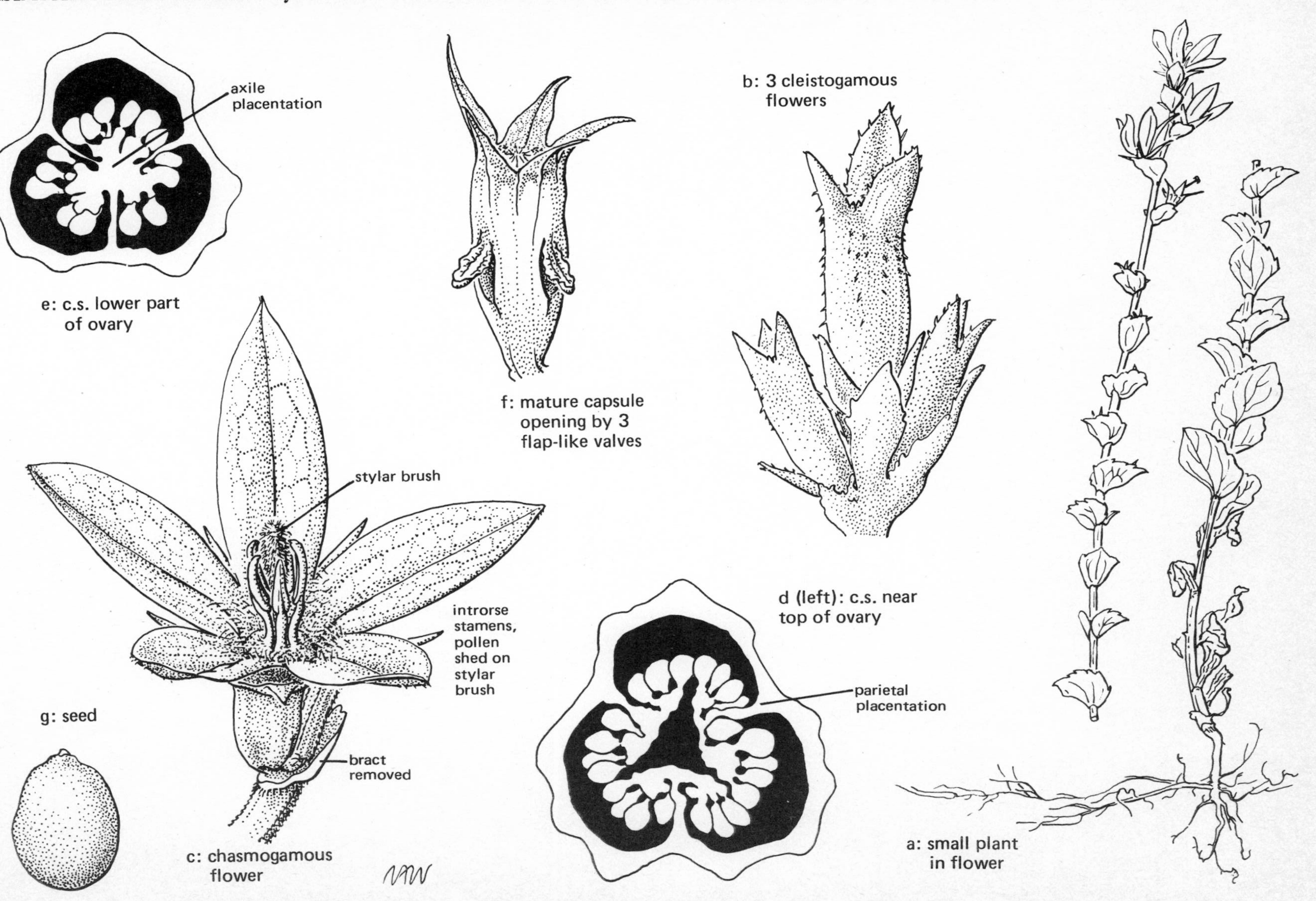

a: small plant in flower

b: 3 cleistogamous flowers

d (left): c.s. near top of ovary

parietal placentation

f: mature capsule opening by 3 flap-like valves

e: c.s. lower part of ovary

axile placentation

c: chasmogamous flower

stylar brush

introrse stamens, pollen shed on stylar brush

bract removed

g: seed

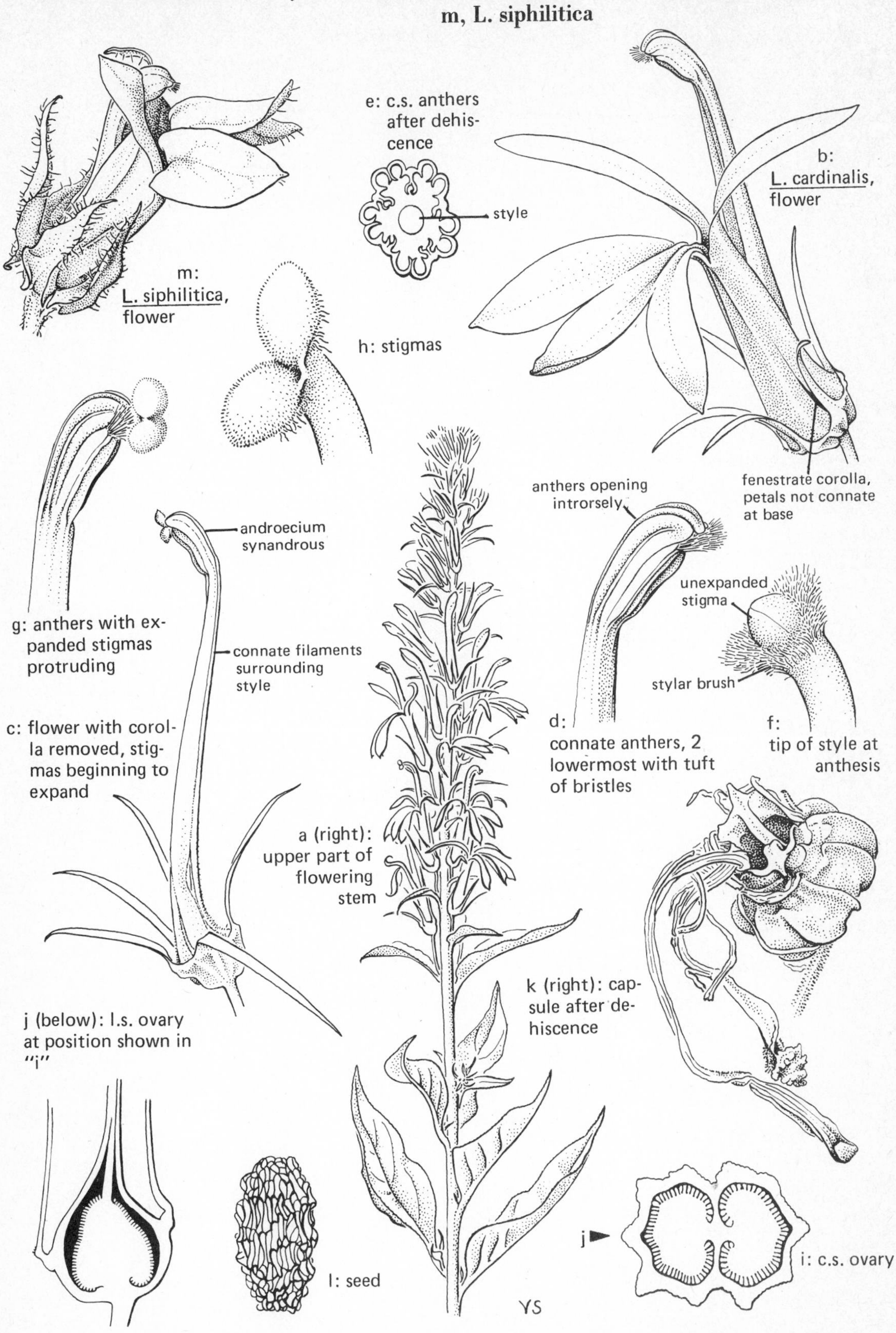

e: c.s. anthers
after dehis-
cence

style

b:
L. cardinalis,
flower

m:
L. siphilitica,
flower

h: stigmas

anthers opening
introrsely

fenestrate corolla,
petals not connate
at base

androecium
synandrous

unexpanded
stigma

g: anthers with ex-
panded stigmas
protruding

connate filaments
surrounding
style

stylar brush

c: flower with corol-
la removed, stig-
mas beginning to
expand

d:
connate anthers, 2
lowermost with tuft
of bristles

f:
tip of style at
anthesis

a (right):
upper part of
flowering
stem

k (right): cap-
sule after de-
hiscence

j (below): l.s. ovary
at position shown in
"i"

l: seed

j ▶

i: c.s. ovary

YS

ASTERACEAE (COMPOSITAE): Inflorescence, florets, and stamens

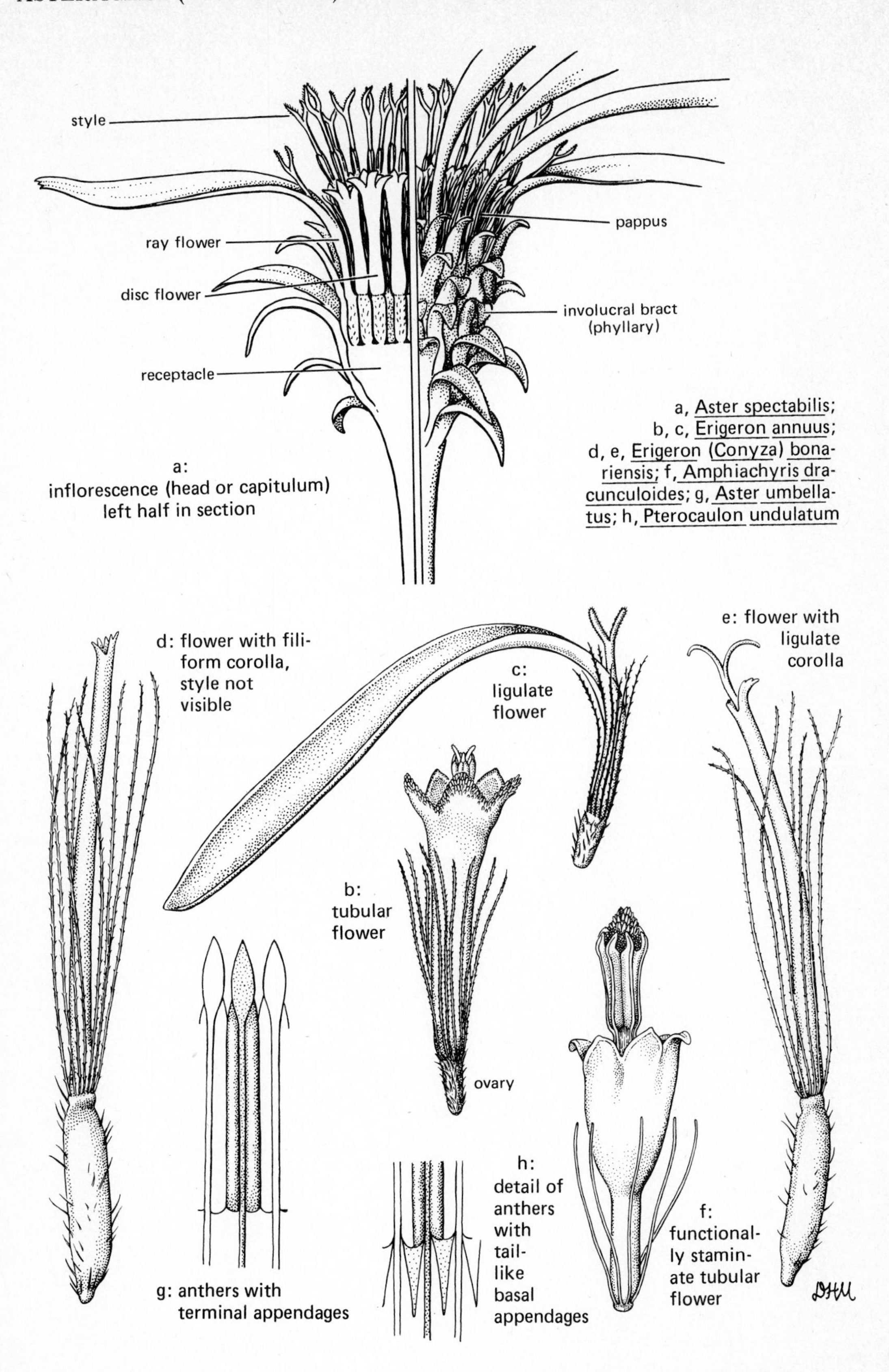

style

pappus

ray flower

disc flower

involucral bract
(phyllary)

receptacle

a:
inflorescence (head or capitulum)
left half in section

a, Aster spectabilis;
b, c, Erigeron annuus;
d, e, Erigeron (Conyza) bona-
riensis; f, Amphiachyris dra-
cunculoides; g, Aster umbella-
tus; h, Pterocaulon undulatum

d: flower with fili-
form corolla,
style not
visible

c:
ligulate
flower

e: flower with
ligulate
corolla

b:
tubular
flower

ovary

g: anthers with
terminal appendages

h:
detail of
anthers
with
tail-
like
basal
appendages

f:
functional-
ly stamin-
ate tubular
flower

115

ASTERACEAE (COMPOSITAE): a-l, characteristic stylar types of genera of eight tribes

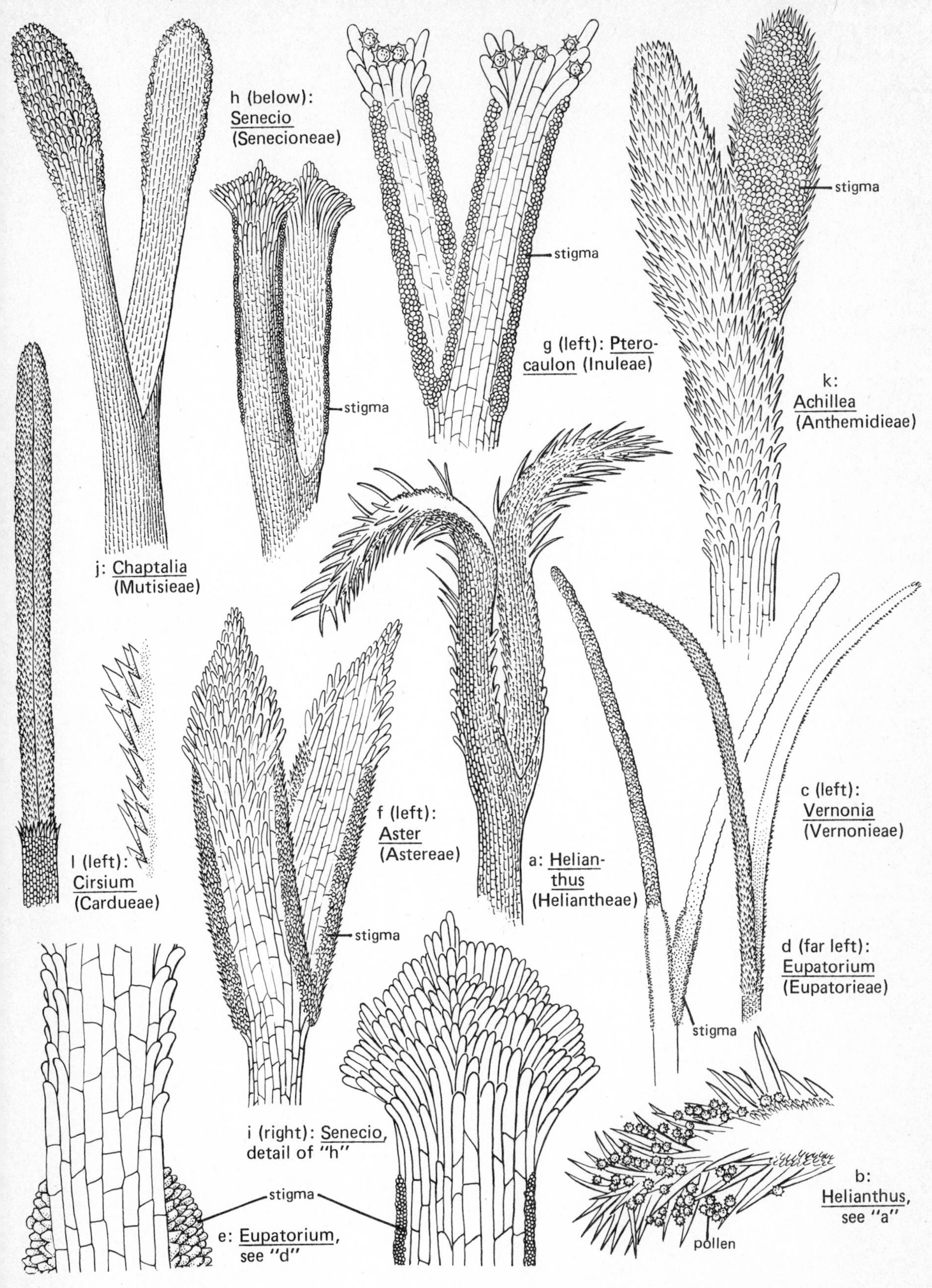

h (below): Senecio (Senecioneae)

g (left): Pterocaulon (Inuleae)

stigma

k: Achillea (Anthemidieae)

stigma

j: Chaptalia (Mutisieae)

c (left): Vernonia (Vernonieae)

l (left): Cirsium (Cardueae)

f (left): Aster (Astereae)

stigma

a: Helianthus (Heliantheae)

d (far left): Eupatorium (Eupatorieae)

stigma

i (right): Senecio, detail of "h"

stigma

e: Eupatorium, see "d"

b: Helianthus, see "a"

pollen

ASTERACEAE (COMPOSITAE) tribe HELIANTHEAE: Helianthus. a-j, H. tuberosus; k, l, H. annuus

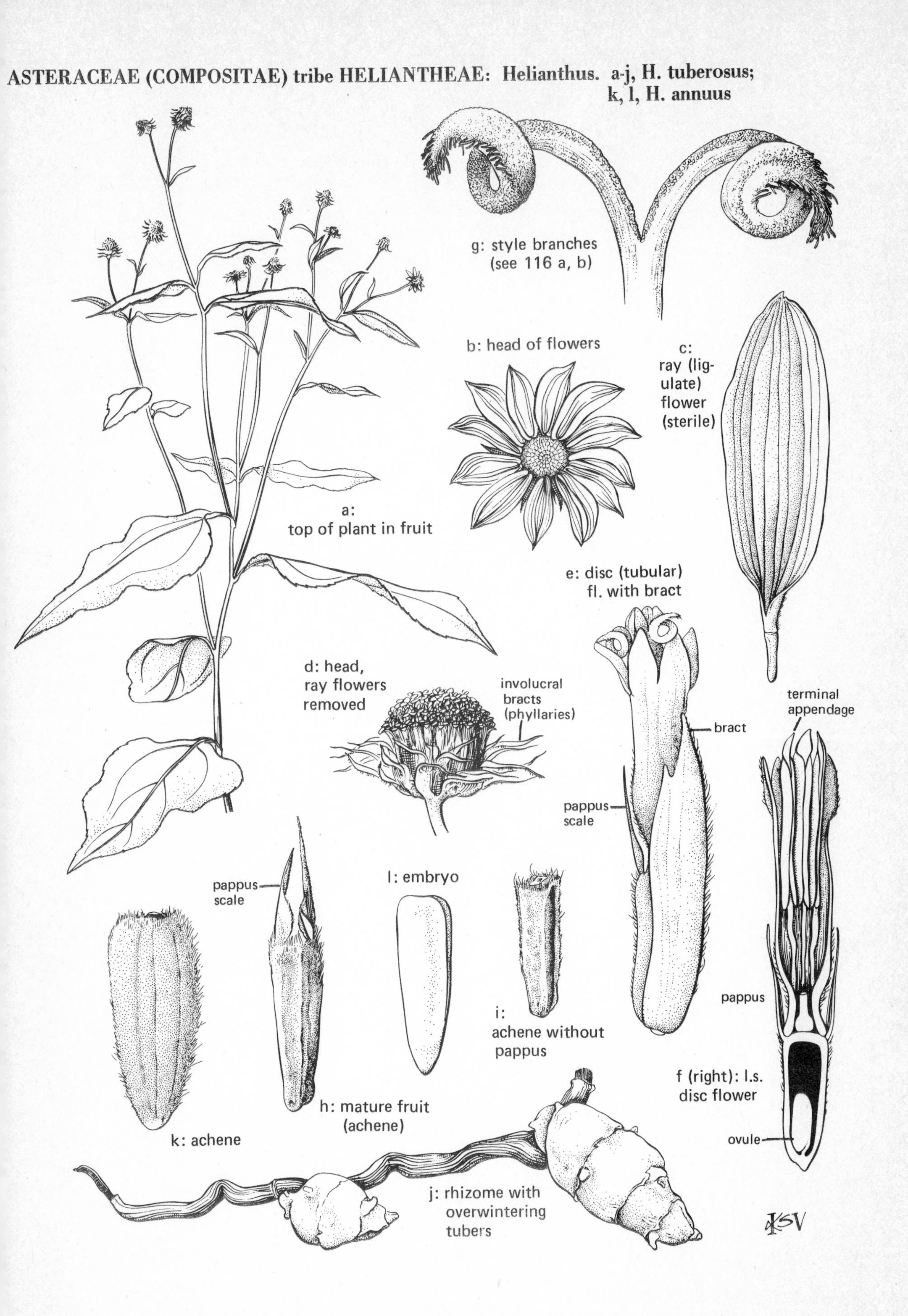

g: style branches
(see 116 a, b)

b: head of flowers

c: ray (lig-ulate) flower (sterile)

a: top of plant in fruit

e: disc (tubular) fl. with bract

d: head, ray flowers removed

involucral bracts (phyllaries)

terminal appendage

bract

pappus scale

pappus scale

l: embryo

pappus

i: achene without pappus

f (right): l.s. disc flower

h: mature fruit (achene)

k: achene

ovule

j: rhizome with overwintering tubers

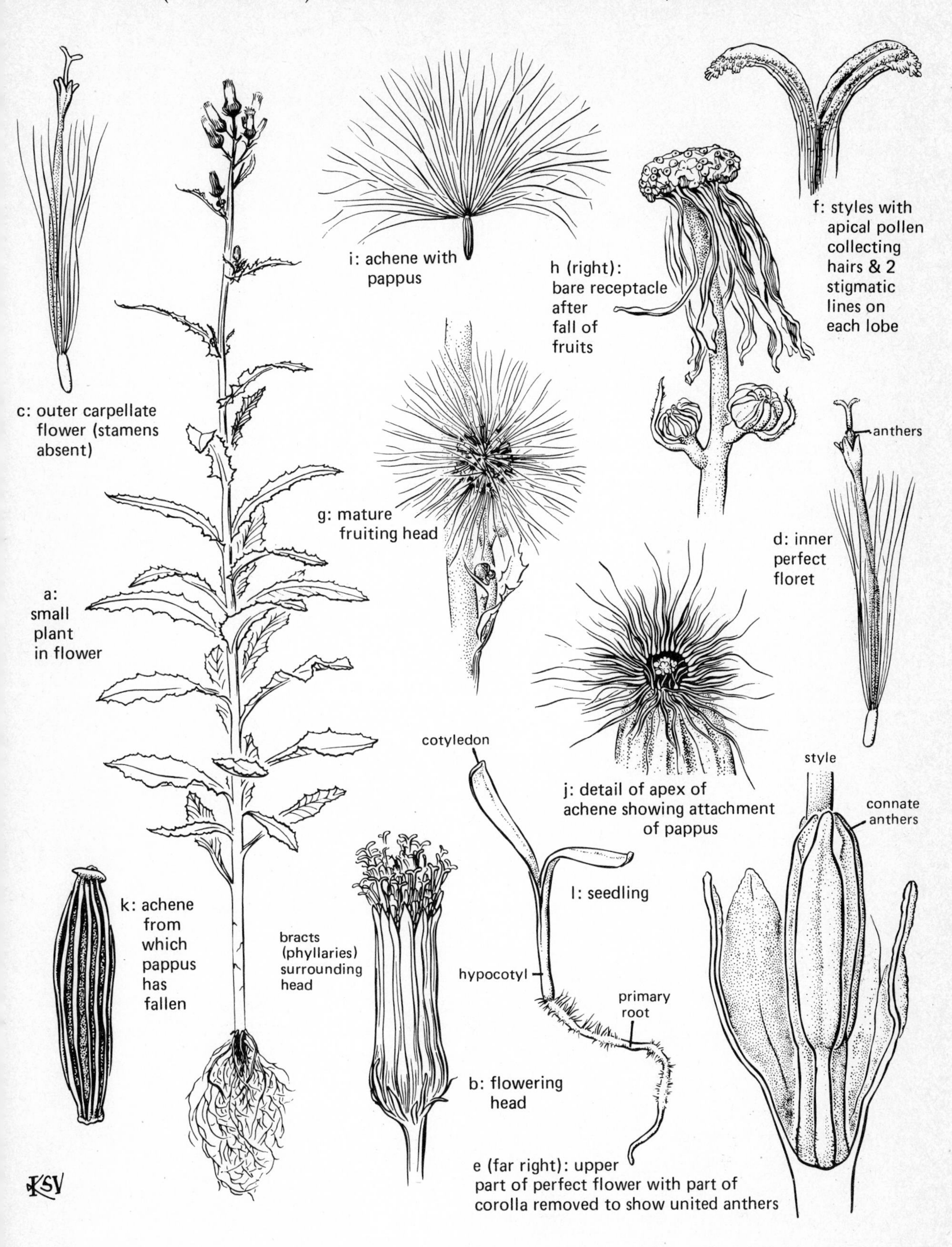

c: outer carpellate
flower (stamens
absent)

i: achene with
pappus

h (right):
bare receptacle
after
fall of
fruits

f: styles with
apical pollen
collecting
hairs & 2
stigmatic
lines on
each lobe

g: mature
fruiting head

a:
small
plant
in flower

anthers

d: inner
perfect
floret

cotyledon

j: detail of apex of
achene showing attachment
of pappus

style

connate
anthers

k: achene
from
which
pappus
has
fallen

l: seedling

bracts
(phyllaries)
surrounding
head

hypocotyl

primary
root

b: flowering
head

e (far right): upper
part of perfect flower with part of
corolla removed to show united anthers

KSV

STERACEAE (COMPOSITAE) tribe HELIANTHEAE: Ambrosia. a-i, A. artemisiifolia; j, A. bidentata; k, A. pilostachya; l, A. trifida

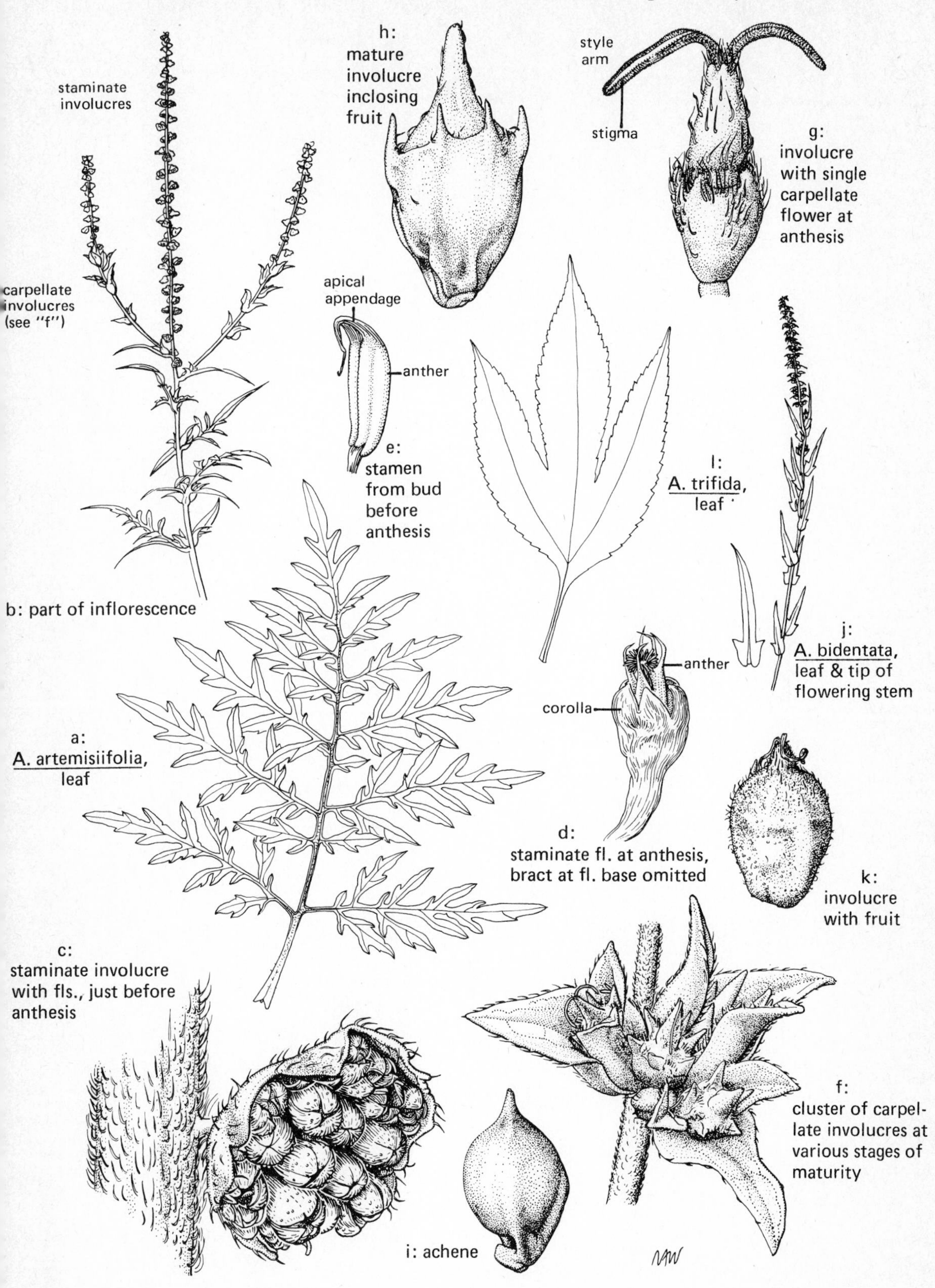

staminate involucres

carpellate involucres (see "f")

b: part of inflorescence

a: A. artemisiifolia, leaf

c: staminate involucre with fls., just before anthesis

h: mature involucre inclosing fruit

apical appendage

anther

e: stamen from bud before anthesis

d: staminate fl. at anthesis, bract at fl. base omitted

corolla

anther

i: achene

style arm

stigma

g: involucre with single carpellate flower at anthesis

l: A. trifida, leaf

j: A. bidentata, leaf & tip of flowering stem

k: involucre with fruit

f: cluster of carpellate involucres at various stages of maturity

119

ASTERACEAE (COMPOSITAE) tribe LACTUCEAE: Cichorium.   a-l, C. intybus

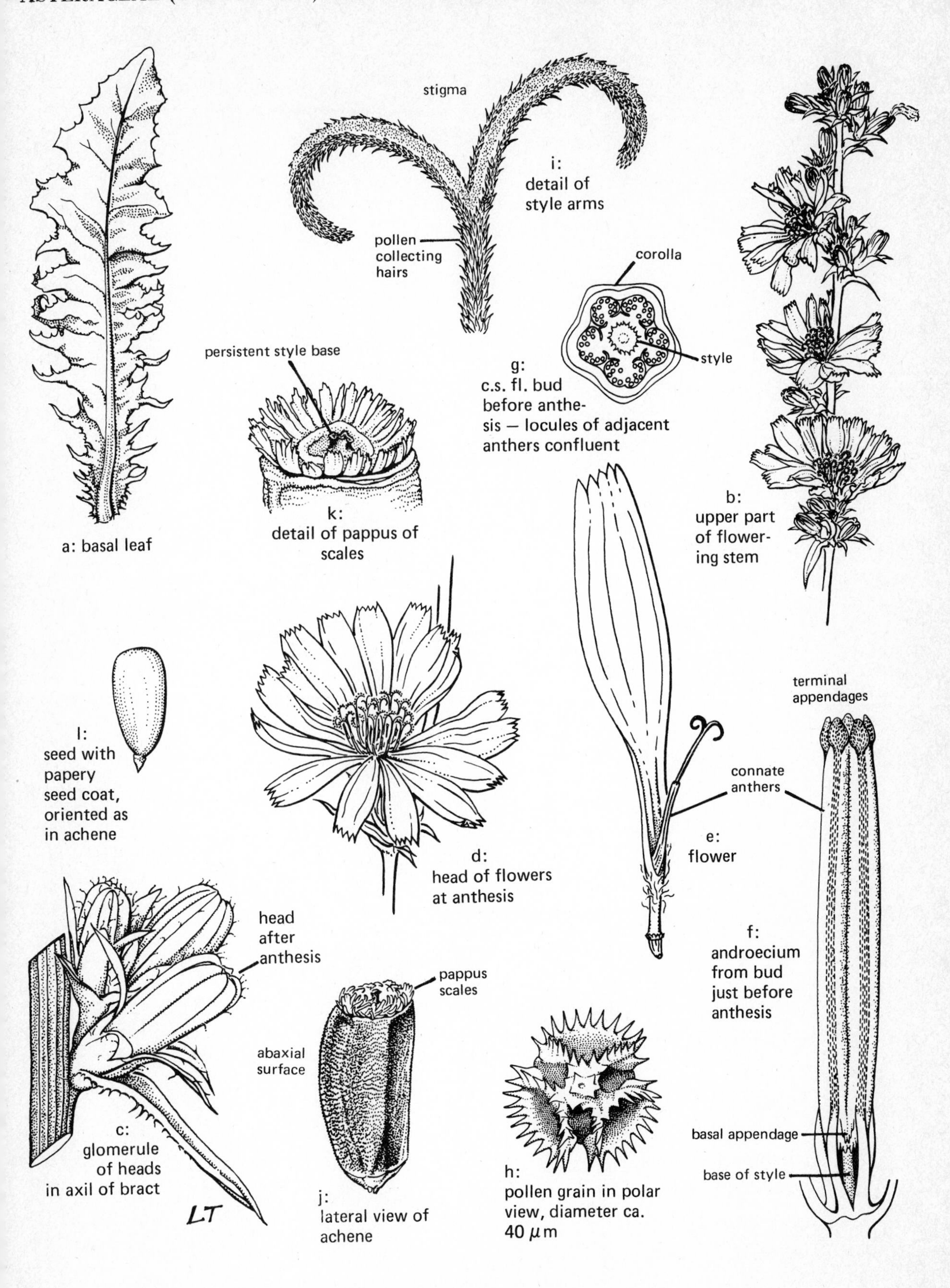

stigma

i:
detail of
style arms

pollen
collecting
hairs

corolla

g:
c.s. fl. bud
before anthe-
sis — locules of adjacent
anthers confluent

style

persistent style base

k:
detail of pappus of
scales

b:
upper part
of flower-
ing stem

a: basal leaf

terminal
appendages

l:
seed with
papery
seed coat,
oriented as
in achene

connate
anthers

e:
flower

d:
head of flowers
at anthesis

f:
androecium
from bud
just before
anthesis

head
after
anthesis

pappus
scales

abaxial
surface

basal appendage

base of style

c:
glomerule
of heads
in axil of bract

LT

j:
lateral view of
achene

h:
pollen grain in polar
view, diameter ca.
40 μm

INDEX TO ILLUSTRATIONS

FAMILIES

GENERA

Generic names in parentheses are those of Brassicaceae (Cruciferae), Apiaceae (Umbelliferae), or of Asteraceae (Compositae) that are illustrated only by infructescences (42, 43, 79) or by florets, stamens, or styles (115, 116).

75 76 77 9 8 7 6 5 4 3 2